T0220141

Introduction to Malliavin Calculus

This textbook offers a compact introductory course on Malliavin calculus, an active
and powerful area of research. It covers recent applications including density
formulas, regularity of probability laws, central and noncentral limit theorems for
Gaussian functionals, convergence of densities, and noncentral limit theorems for the
local time of Brownian motion. The book also includes self-contained presentations of
Brownian motion and stochastic calculus as well as of Lévy processes and stochastic
calculus for jump processes. Accessible to nonexperts, the book can be used by
graduate students and researchers to develop their mastery of the core techniques
necessary for further study.

DAVID NUALART is the Black–Babcock Distinguished Professor in the Department of
Mathematics of Kansas University. He has published around 300 scientific articles in
the field of probability and stochastic processes, and he is the author of the
fundamental monograph *The Malliavin Calculus and Related Topics*. He has served on
the editorial board of leading journals in probability, and from 2006 to 2008 was the
editor-in-chief of *Electronic Communications in Probability*. He was elected Fellow of
the US Institute of Mathematical Statistics in 1997 and received the Higuchi Award on
Basic Sciences in 2015.

EULALIA NUALART is an Associate Professor at Universitat Pompeu Fabra and a
Barcelona GSE Affiliated Professor. She is also the Deputy Director of the Barcelona
GSE Master Program in Economics. Her research interests include stochastic analysis,
Malliavin calculus, fractional Brownian motion, and Lévy processes. She has
publications in journals such as *Stochastic Processes and their Applications*, *Annals of
Probability*, and *Journal of Functional Analysis*. In 2013 she was awarded a Marie
Curie Career Integration Grant.

INSTITUTE OF MATHEMATICAL STATISTICS
TEXTBOOKS

Editorial Board
N. Reid (University of Toronto)
R. van Handel (Princeton University)
S. Holmes (Stanford University)
X. He (University of Michigan)

IMS Textbooks give introductory accounts of topics of current concern suitable for advanced courses at master's level, for doctoral students, and for individual study. They are typically shorter than a fully developed textbook, often arising from material created for a topical course. Lengths of 100–290 pages are envisaged. The books typically contain exercises.

Other books in the series

Introduction to Malliavin Calculus

DAVID NUALART
University of Kansas

EULALIA NUALART
Universitat Pompeu Fabra, Barcelona

CAMBRIDGE
UNIVERSITY PRESS

CAMBRIDGE
UNIVERSITY PRESS

Shaftesbury Road, Cambridge CB2 8EA, United Kingdom

One Liberty Plaza, 20th Floor, New York, NY 10006, USA

477 Williamstown Road, Port Melbourne, VIC 3207, Australia

314–321, 3rd Floor, Plot 3, Splendor Forum, Jasola District Centre, New Delhi – 110025, India

103 Penang Road, #05–06/07, Visioncrest Commercial, Singapore 238467

Cambridge University Press is part of Cambridge University Press & Assessment, a department of the University of Cambridge.

We share the University's mission to contribute to society through the pursuit of education, learning and research at the highest international levels of excellence.

www.cambridge.org
Information on this title: www.cambridge.org/9781107611986

DOI: 10.1017/9781139856485

© David Nualart and Eulalia Nualart 2018

This publication is in copyright. Subject to statutory exception and to the provisions of relevant collective licensing agreements, no reproduction of any part may take place without the written permission of Cambridge University Press & Assessment.

First published 2018

A catalogue record for this publication is available from the British Library

Library of Congress Cataloging-in-Publication data
Names: Nualart, David, 1951– author. | Nualart, Eulalia, author.
Title: Introduction to Malliavin calculus / David Nualart (University of Kansas), Eulalia Nualart (Universitat Pompeu Fabra, Barcelona).
Description: Cambridge : Cambridge University Press, [2018] |
Series: Institute of Mathematical Statistics textbooks |
Includes bibliographical references and index.
Identifiers: LCCN 2018013735 | ISBN 9781107039124 (alk. paper)
Subjects: LCSH: Malliavin calculus. | Stochastic analysis. | Derivatives (Mathematics) | Calculus of variations.
Classification: LCC QA174.2 .N83 2018 | DDC 519.2/3–dc23
LC record available at https://lccn.loc.gov/2018013735

ISBN   978-1-107-03912-4   Hardback
ISBN   978-1-107-61198-6   Paperback

Cambridge University Press & Assessment has no responsibility for the persistence or accuracy of URLs for external or third-party internet websites referred to in this publication and does not guarantee that any content on such websites is, or will remain, accurate or appropriate.

To my wife, Maria Pilar
To my daughter, Juliette

Contents

Preface

This textbook provides an introductory course on Malliavin calculus intended to prepare the interested reader for further study of existing monographs on the subject such as Bichteler *et al.* (1987), Malliavin (1991), Sanz-Solé (2005), Malliavin and Thalmaier (2005), Nualart (2006), Di Nunno *et al.* (2009), Nourdin and Peccati (2012), and Ishikawa (2016), among others. Moreover, it contains recent applications of Malliavin calculus, including density formulas, central limit theorems for functionals of Gaussian processes, theorems on the convergence of densities, noncentral limit theorems, and Malliavin calculus for jump processes. Recommended prior knowledge would be an advanced probability course that includes laws of large numbers and central limit theorems, martingales, and Markov processes.

The Malliavin calculus is an infinite-dimensional differential calculus on Wiener space, first introduced by Paul Malliavin in the 1970s with the aim of giving a probabilistic proof of Hörmander's hypoellipticity theorem; see Malliavin (1978a, b, c). The theory was further developed, see e.g. Shigekawa (1980), Bismut (1981), Stroock (1981a, b), and Ikeda and Watanabe (1984), and since then many new applications have appeared.

Chapters 1 and 2 give an introduction to stochastic calculus with respect to Brownian motion, as developed by Itô (1944). The purpose of this calculus is to construct stochastic integrals for adapted and square integrable processes and to develop a change-of-variable formula.

Chapters 3, 4, and 5 present the main operators of the Malliavin calculus, which are the derivative, the divergence, the generator of the Ornstein–Uhlenbeck semigroup, and the corresponding Sobolev norms. In Chapter 4, multiple stochastic integrals are constructed following Itô (1951), and the orthogonal decomposition of square integrable random variables due to Wiener (1938) is derived. These concepts play a key role in the development of further properties of the Malliavin calculus operators. In particular, Chapter 5 contains an integration-by-parts formula that relates the three operators, which is crucial for applications. In particular, it allows us to prove a density formula due to Nourdin and Viens (2009).

Chapters 6, 7, and 8 are devoted to different applications of the Malliavin calculus for Brownian motion. Chapter 6 presents two different stochastic integral representations: the first is the well-known Clark–Ocone formula, and the second uses the inverse of the Ornstein–Ulhenbeck generator. We present, as a consequence of the Clark–Ocone formula, a central limit theorem for the modulus of continuity of the local time of Brownian motion, proved by Hu and Nualart (2009). As an application of the second representation formula, we show how to derive tightness in the asymptotic behavior of the self-intersection local time of fractional Brownian motion, following Hu and Nualart (2005) and Jaramillo and Nualart (2018). In Chapter 7 we develop the Malliavin calculus to derive explicit formulas for the densities of random variables and criteria for their regularity. We apply these criteria to the proof of Hörmander's hypoellipticity theorem. Chapter 8 presents an application of Malliavin calculus, combined with Stein's method, to normal approximations.

Chapters 9, 10, and 11 develop Malliavin calculus for Poisson random measures. Specifically, Chapter 9 introduces stochastic integration for jump processes, as well as the Wiener chaos decomposition of a Poisson random measure. Then the Malliavin calculus is developed in two different directions. In Chapter 10 we introduce the three Malliavin operators and their Sobolev norms using the Wiener chaos decomposition. As an application, we present the Clark–Ocone formula and Stein's method for Poisson functionals. In Chapter 11 we use the theory of cylindrical functionals to introduce the derivative and divergence operators. This approach allows us to obtain a criterion for the existence of densities, which we apply to diffusions with jumps.

Finally, in the appendix we review basic results on stochastic processes that are used throughout the book.

1

Brownian Motion

In this chapter we introduce Brownian motion and study several aspects of this stochastic process, including the regularity of sample paths, quadratic variation, Wiener stochastic integrals, martingales, Markov properties, hitting times, and the reflection principle.

1.1 Preliminaries and Notation

Throughout this book we will denote by $(\Omega, \mathcal{F}, P)$ a probability space, where Ω is a sample space, $\mathcal{F}$ is a σ-algebra of subsets of Ω, and P is a σ-additive probability measure on $(\Omega, \mathcal{F})$. If X is an integrable or nonnegative random variable on $(\Omega, \mathcal{F}, P)$, we denote by $E(X)$ its expectation. For any $p \geq 1$, we denote by $L^p(\Omega)$ the space of random variables on $(\Omega, \mathcal{F}, P)$ such that the norm

$$\|X\|_p := (E(|X|^p))^{1/p}$$

is finite.

For any integers $k, n \geq 1$ we denote by $C_b^k(\mathbb{R}^n)$ the space of k-times continuously differentiable functions $f : \mathbb{R}^n \to \mathbb{R}$, such that f and all its partial derivatives of order up to k are bounded. We also denote by $C_0^k(\mathbb{R}^n)$ the subspace of functions in $C_b^k(\mathbb{R}^n)$ that have compact support. Moreover, $C_p^\infty(\mathbb{R}^n)$ is the space of infinitely differentiable functions on $\mathbb{R}^n$ that have at most polynomial growth together with their partial derivatives, $C_b^\infty(\mathbb{R}^n)$ is the subspace of functions in $C_p^\infty(\mathbb{R}^n)$ that are bounded together with their partial derivatives, and $C_0^\infty(\mathbb{R}^n)$ is the space of infinitely differentiable functions with compact support.

1.2 Definition and Basic Properties

Brownian motion was named by Einstein (1905) after the botanist Robert Brown (1828), who observed in a microscope the complex and erratic motion of grains of pollen suspended in water. Brownian motion was then rigorously defined and studied by Wiener (1923); this is why it is also called the Wiener process. For extended expositions about Brownian motion see Revuz and Yor (1999), Mörters and Peres (2010), Durrett (2010), Bass (2011), and Baudoin (2014).

The mathematical definition of Brownian motion is the following.

Definition 1.2.1 A real-valued stochastic process $B = (B_t)_{t\geq0}$ defined on a probability space $(\Omega, \mathcal{F}, P)$ is called a *Brownian motion* if it satisfies the following conditions:

(i) Almost surely $B_0 = 0$.

(ii) For all $0 \leq t_1 < \cdots < t_n$ the increments $B_{t_n} - B_{t_{n-1}}, \ldots, B_{t_2} - B_{t_1}$ are independent random variables.

(iii) If $0 \leq s < t$, the increment $B_t - B_s$ is a Gaussian random variable with mean zero and variance $t - s$.

(iv) With probability one, the map $t \to B_t$ is continuous.

More generally, a d-dimensional Brownian motion is defined as an $\mathbb{R}^d$-valued stochastic process $B = (B_t)_{t\geq0}$, $B_t = (B_t^1, \ldots, B_t^d)$, where $B^1, \ldots, B^d$ are d independent Brownian motions.

We will sometimes consider a Brownian motion on a finite time interval $[0, T]$, which is defined in the same way.

Proposition 1.2.2 *Properties* (i), (ii), *and* (iii) *are equivalent to saying that B is a Gaussian process with mean zero and covariance function*

$$\Gamma(s, t) = \min(s, t). \tag{1.1}$$

Proof Suppose that (i), (ii), and (iii) hold. The probability distribution of the random vector $(B_{t_1}, \ldots, B_{t_n})$, for $0 < t_1 < \cdots < t_n$, is normal because this vector is a linear transformation of the vector

$$(B_{t_1}, B_{t_2} - B_{t_1}, \ldots, B_{t_n} - B_{t_{n-1}}),$$

which has a normal distribution because its components are independent and normal. The mean $m(t)$ and the covariance function $\Gamma(s, t)$ are given by

$$m(t) = E(B_t) = 0,$$
$$\Gamma(s, t) = E(B_s B_t) = E(B_s(B_t - B_s + B_s))$$
$$= E(B_s(B_t - B_s)) + E(B_s^2) = s = \min(s, t),$$

if $s \leq t$. The converse is also easy to show. □

The existence of Brownian motion can be proved in different ways.

(1) The function $\Gamma(s, t) = \min(s, t)$ is symmetric and nonnegative definite because it can be written as

$$\min(s, t) = \int_0^\infty \mathbf{1}_{[0,s]}(r)\mathbf{1}_{[0,t]}(r)dr.$$

Then, for any integer $n \geq 1$ and real numbers $a_1, \ldots, a_n$,

$$\sum_{i,j=1}^n a_i a_j \min(t_i, t_j) = \sum_{i,j=1}^n a_i a_j \int_0^\infty \mathbf{1}_{[0,t_i]}(r)\mathbf{1}_{[0,t_j]}(r)dr$$

$$= \int_0^\infty \left(\sum_{i=1}^n a_i \mathbf{1}_{[0,t_i]}(r)\right)^2 dr \geq 0.$$

Therefore, by Kolmogorov's extension theorem (Theorem A.1.1), there exists a Gaussian process with mean zero and covariance function $\min(s, t)$.

Moreover, for any $s \leq t$, the increment $B_t - B_s$ has the normal distribution $N(0, t - s)$. This implies that for any natural number k we have

$$\mathrm{E}\left((B_t - B_s)^{2k}\right) = \frac{(2k)!}{2^k k!}(t - s)^k.$$

Therefore, by Kolmogorov's continuity theorem (Theorem A.4.1), there exists a version of B with Hölder-continuous trajectories of order γ for any $\gamma < (k - 1)/(2k)$ on any interval $[0, T]$. This implies that the paths of this version of the process B are γ-Hölder continuous on $[0, T]$ for any $\gamma < 1/2$ and $T > 0$.

(2) Brownian motion can also be constructed as a Fourier series with random coefficients. Fix $T > 0$ and suppose that $(e_n)_{n \geq 0}$ is an orthonormal basis of the Hilbert space $L^2([0, T])$. Suppose that $(Z_n)_{n \geq 0}$ are independent random variables with law $N(0, 1)$. Then, the random series

$$\sum_{n=0}^\infty Z_n \int_0^t e_n(r)dr \tag{1.2}$$

converges in $L^2(\Omega)$ to a mean-zero Gaussian process $B = (B_t)_{t \in [0,T]}$ with covariance function (1.1). In fact, for any $s, t \in [0, T]$,

$$\mathrm{E}\left(\left(\sum_{n=0}^{N} Z_n \int_0^t e_n(r)dr\right)\left(\sum_{n=0}^{N} Z_n \int_0^s e_n(r)dr\right)\right)$$

$$= \sum_{n=0}^{N}\left(\int_0^t e_n(r)dr\right)\left(\int_0^s e_n(r)dr\right)$$

$$= \sum_{n=0}^{N} \langle \mathbf{1}_{[0,t]}, e_n \rangle_{L^2([0,T])} \langle \mathbf{1}_{[0,s]}, e_n \rangle_{L^2([0,T])},$$

which converges as $N \to \infty$ to

$$\langle \mathbf{1}_{[0,t]}, \mathbf{1}_{[0,s]} \rangle_{L^2([0,T])} = \min(s, t).$$

The convergence of the series (1.2) is uniform in $[0, T]$ almost surely; that is, as N tends to infinity,

$$\sup_{0 \le t \le T}\left|\sum_{n=0}^{N} Z_n \int_0^t e_n(r)dr - B_t\right| \xrightarrow{\text{a.s.}} 0. \tag{1.3}$$

The fact that the process B has continuous trajectories almost surely is a consequence of (1.3). We refer to Itô and Nisio (1968) for a proof of (1.3).

Once we have constructed the Brownian motion on an interval $[0, T]$, we can build a Brownian motion on $\mathbb{R}_+$ by considering a sequence of independent Brownian motions $B^{(n)}$ on $[0, T]$, $n \ge 1$, and setting

$$B_t = B_T^{(n-1)} + B_{t-(n-1)T}^{(n)}, \quad (n-1)T \le t \le nT,$$

with the convention $B_T^{(0)} = 0$.

In particular, if we take a basis formed by the trigonometric functions, $e_n(t) = (1/\sqrt{\pi})\cos(nt/2)$ for $n \ge 1$ and $e_0(t) = 1/\sqrt{2\pi}$, on the interval $[0, 2\pi]$, we obtain the Paley–Wiener representation of Brownian motion:

$$B_t = Z_0 \frac{t}{\sqrt{2\pi}} + \frac{2}{\sqrt{\pi}} \sum_{n=1}^{\infty} Z_n \frac{\sin(nt/2)}{n}, \quad t \in [0, 2\pi]. \tag{1.4}$$

The proof of the construction of Brownian motion in this particular case can be found in Bass (2011, Theorem 6.1).

(3) Brownian motion can also be regarded as the limit in distribution of a symmetric random walk. Indeed, fix a time interval $[0, T]$. Consider n independent and identically distributed random variables $\xi_1, \ldots, \xi_n$ with mean zero and variance T/n. Define the partial sums

$$R_k = \xi_1 + \cdots + \xi_k, \quad k = 1, \ldots, n.$$

By the central limit theorem the sequence R_n converges in distribution, as n tends to infinity, to the normal distribution $N(0, T)$.

Consider the continuous stochastic process $S_n(t)$ defined by linear interpolation from the values

$$S_n\left(\frac{kT}{n}\right) = R_k, \quad k = 0, \ldots, n.$$

Then, a functional version of the central limit theorem, known as the Donsker invariance principle, says that the sequence of stochastic processes $S_n(t)$ converges in law to Brownian motion on $[0, T]$. This means that, for any continuous and bounded function $\varphi \colon C([0, T]) \to \mathbb{R}$, we have

$$E(\varphi(S_n)) \to E(\varphi(B)),$$

as n tends to infinity.

Basic properties of Brownian motion are (see Exercises 1.5–1.8):

1. *Self-similarity* For any $a > 0$, the process $(a^{-1/2}B_{at})_{t \geq 0}$ is a Brownian motion.
2. For any $h > 0$, the process $(B_{t+h} - B_h)_{t \geq 0}$ is a Brownian motion.
3. The process $(-B_t)_{t \geq 0}$ is a Brownian motion.
4. Almost surely $\lim_{t \to \infty} B_t/t = 0$, and the process

$$X_t = \begin{cases} tB_{1/t} & \text{if } t > 0, \\ 0 & \text{if } t = 0, \end{cases}$$

is a Brownian motion.

Remark 1.2.3 As we have seen, the trajectories of Brownian motion on an interval $[0, T]$ are Hölder continuous of order γ for any $\gamma < \frac{1}{2}$. However, the trajectories are not Hölder continuous of order $\frac{1}{2}$. More precisely, the following property holds (see Exercise 1.9):

$$P\left(\sup_{s,t \in [0,1]} \frac{|B_t - B_s|}{\sqrt{|t - s|}} = +\infty\right) = 1.$$

The *exact modulus of continuity* of Brownian motion was obtained by Lévy (1937):

$$\limsup_{\delta \downarrow 0} \sup_{s,t \in [0,1], |t-s| < \delta} \frac{|B_t - B_s|}{\sqrt{2|t - s| \log |t - s|}} = 1, \quad \text{a.s.}$$

Lévy's proof can be found in Mörters and Peres (2010, Theorem 1.14). In contrast, the behavior at a single point is given by the *law of the iterated logarithm*, due to Khinchin (1933):

$$\limsup_{t \downarrow s} \frac{|B_t - B_s|}{\sqrt{2|t - s| \log \log |t - s|}} = 1, \quad \text{a.s.}$$

for any $s \geq 0$. See also Mörters and Peres (2010, Corollary 5.3) and Bass (2011, Theorem 7.2).

Brownian motion satisfies $E(|B_t - B_s|^2) = t - s$ for all $s \leq t$. This means that when $t - s$ is small, $B_t - B_s$ is of order $\sqrt{t - s}$ and $(B_t - B_s)^2$ is of order $t - s$. Moreover, the quadratic variation of a Brownian motion on $[0, t]$ equals t in $L^2(\Omega)$, as is proved in the following proposition.

Proposition 1.2.4 *Fix a time interval $[0, t]$ and consider the following subdivision π of this interval:*

$$0 = t_0 < t_1 < \cdots < t_n = t.$$

The norm of the subdivision π is defined as $|\pi| = \max_{0 \leq j \leq n-1}(t_{j+1} - t_j)$. The following convergence holds in $L^2(\Omega)$:

$$\lim_{|\pi| \to 0} \sum_{j=0}^{n-1} (B_{t_{j+1}} - B_{t_j})^2 = t. \tag{1.5}$$

Proof Set $\xi_j = (B_{t_{j+1}} - B_{t_j})^2 - (t_{j+1} - t_j)$. The random variables ξ_j are independent and centered. Thus,

$$E\left(\left(\sum_{j=0}^{n-1}(B_{t_{j+1}} - B_{t_j})^2 - t\right)^2\right) = E\left(\left(\sum_{j=0}^{n-1}\xi_j\right)^2\right) = \sum_{j=0}^{n-1} E\left(\xi_j^2\right)$$

$$= \sum_{j=0}^{n-1}\left(3(t_{j+1} - t_j)^2 - 2(t_{j+1} - t_j)^2 + (t_{j+1} - t_j)^2\right)$$

$$= 2\sum_{j=0}^{n-1}(t_{j+1} - t_j)^2 \leq 2t|\pi| \xrightarrow{|\pi| \to 0} 0,$$

which proves the result. □

As a consequence, we have the following result.

Proposition 1.2.5 *The total variation of Brownian motion on an interval $[0, t]$, defined by*

$$V = \sup_{\pi} \sum_{j=0}^{n-1} |B_{t_{j+1}} - B_{t_j}|,$$

where $\pi = \{0 = t_0 < t_1 < \cdots < t_n\}$, is infinite with probability one.

Proof Using the continuity of the trajectories of Brownian motion, we have

$$\sum_{j=1}^{n-1}(B_{t_{j+1}} - B_{t_j})^2 \le \sup_j |B_{t_{j+1}} - B_{t_j}| \left(\sum_{j=0}^{n-1} |B_{t_{j+1}} - B_{t_j}| \right)$$

$$\le V \sup_j |B_{t_{j+1}} - B_{t_j}| \overset{|\pi| \to 0}{\longrightarrow} 0$$

if $V < \infty$, which contradicts the fact that $\sum_{j=0}^{n-1}(B_{t_{j+1}} - B_{t_j})^2$ converges in mean square to t as $|\pi| \to 0$. Therefore, $P(V < \infty) = 0$. □

Finally, the trajectories of B are almost surely nowhere differentiable. The first proof of this fact is due to Paley *et al.* (1933). Another proof, by Dvoretzky *et al.* (1961), is given in Durrett (2010, Theorem 8.1.6) and Mörters and Peres (2010, Theorem 1.27).

1.3 Wiener Integral

We next define the integral of square integrable functions with respect to Brownian motion, known as the Wiener integral.

We consider the set $\mathcal{E}_0$ of step functions

$$\varphi_t = \sum_{j=0}^{n-1} a_j \mathbf{1}_{(t_j, t_{j+1}]}(t), \quad t \ge 0, \tag{1.6}$$

where $n \ge 1$ is an integer, $a_0, \ldots, a_{n-1} \in \mathbb{R}$, and $0 = t_0 < \cdots < t_n$. The Wiener integral of a step function $\varphi \in \mathcal{E}_0$ of the form (1.6) is defined by

$$\int_0^\infty \varphi_t dB_t = \sum_{j=0}^{n-1} a_j (B_{t_{j+1}} - B_{t_j}).$$

The mapping $\varphi \to \int_0^\infty \varphi_t dB_t$ from $\mathcal{E}_0 \subset L^2(\mathbb{R}_+)$ to $L^2(\Omega)$ is linear and isometric:

$$E\left(\left(\int_0^\infty \varphi_t dB_t \right)^2 \right) = \sum_{j=0}^{n-1} a_j^2 (t_{j+1} - t_j) = \int_0^\infty \varphi_t^2 dt = \|\varphi\|_{L^2(\mathbb{R}_+)}^2.$$

The space $\mathcal{E}_0$ is a dense subspace of $L^2(\mathbb{R}_+)$. Therefore, the mapping

$$\varphi \to \int_0^\infty \varphi_t dB_t$$

can be extended to a linear isometry between $L^2(\mathbb{R}_+)$ and the Gaussian subspace of $L^2(\Omega)$ spanned by the Brownian motion. The random variable $\int_0^\infty \varphi_t dB_t$ is called the Wiener integral of $\varphi \in L^2(\mathbb{R}_+)$ and is denoted by $B(\varphi)$. Observe that it is a Gaussian random variable with mean zero and variance $\|\varphi\|_{L^2(\mathbb{R}_+)}^2$.

The Wiener integral allows us to view Brownian motion as the cumulative function of a white noise.

Definition 1.3.1   Let D be a Borel subset of $\mathbb{R}^m$. A *white noise* on D is a centered Gaussian family of random variables

$$\{W(A), A \in \mathcal{B}(\mathbb{R}^m), A \subset D, \ell(A) < \infty\},$$

where ℓ denotes the Lebesgue measure, such that

$$E(W(A)W(B)) = \ell(A \cap B).$$

The mapping $\mathbf{1}_A \to W(A)$ can be extended to a linear isometry from $L^2(D)$ to the Gaussian space spanned by W, denoted by

$$\varphi \to \int_D \varphi(x) W(dx).$$

The Brownian motion B defines a white noise on $\mathbb{R}_+$ by setting

$$W(A) = \int_0^\infty \mathbf{1}_A(t) dB_t, \quad A \in \mathcal{B}(\mathbb{R}_+), \ \ell(A) < \infty.$$

Conversely, Brownian motion can be defined from white noise. In fact, if W is a white noise on $\mathbb{R}_+$, the process

$$W_t = W([0, t]), \quad t \geq 0,$$

is a Brownian motion.

The two-parameter extension of Brownian motion is the *Brownian sheet*, which is defined as a real-valued two-parameter Gaussian process $(B_t)_{t \in \mathbb{R}_+^2}$ with mean zero and covariance function

$$\Gamma(s, t) = E(B_s B_t) = \min(s_1, t_1) \min(s_2, t_2), \quad s, t \in \mathbb{R}_+^2.$$

As above, the Brownian sheet can be obtained from white noise. In fact, if W is a white noise on $\mathbb{R}_+^2$, the process

$$W_t = W([0, t_1] \times [0, t_2]), \quad t \in \mathbb{R}_+^2,$$

is a Brownian sheet.

1.4 Wiener Space

Brownian motion can be defined in the canonical probability space $(\Omega, \mathcal{F}, P)$ known as the Wiener space. More precisely:

- Ω is the space of continuous functions $\omega \colon \mathbb{R}_+ \to \mathbb{R}$ vanishing at the origin.
- $\mathcal{F}$ is the Borel σ-field $\mathcal{B}(\Omega)$ for the topology corresponding to uniform convergence on compact sets. One can easily show (see Exercise 1.11) that $\mathcal{F}$ coincides with the σ-field generated by the collection of cylinder sets

$$C = \{\omega \in \Omega : \omega(t_1) \in A_1, \ldots, \omega(t_k) \in A_k\}, \tag{1.7}$$

  for any integer $k \geq 1$, Borel sets $A_1, \ldots, A_k$ in $\mathbb{R}$, and $0 \leq t_1 < \cdots < t_k$.
- P is the Wiener measure. That is, P is defined on a cylinder set of the form (1.7) by

$$P(C) = \int_{A_1 \times \cdots \times A_k} p_{t_1}(x_1) p_{t_2 - t_1}(x_2 - x_1) \cdots p_{t_k - t_{k-1}}(x_k - x_{k-1}) \, dx_1 \cdots dx_k,$$
$$\tag{1.8}$$

where $p_t(x)$ denotes the Gaussian density

$$p_t(x) = (2\pi t)^{-1/2} e^{-x^2/(2t)}, \quad x \in \mathbb{R}, t > 0.$$

The mapping P defined by (1.8) on cylinder sets can be uniquely extended to a probability measure on $\mathcal{F}$. This fact can be proved as a consequence of the existence of Brownian motion on $\mathbb{R}_+$. Finally, the canonical stochastic process defined as $B_t(\omega) = \omega(t)$, $\omega \in \Omega$, $t \geq 0$, is a Brownian motion.

The canonical probability space $(\Omega, \mathcal{F}, P)$ of a d-dimensional Brownian motion can be defined in a similar way.

Further into the text, $(\Omega, \mathcal{F}, P)$ will denote a general probability space, and only in some special cases will we restrict our study to Wiener space.

1.5 Brownian Filtration

Consider a Brownian motion $B = (B_t)_{t \geq 0}$ defined on a probability space $(\Omega, \mathcal{F}, P)$. For any time $t \geq 0$, we define the σ-field $\mathcal{F}_t$ generated by the random variables $(B_s)_{0 \leq s \leq t}$ and the events in $\mathcal{F}$ of probability zero. That is, $\mathcal{F}_t$ is the smallest σ-field that contains the sets of the form

$$\{B_s \in A\} \cup N,$$

where $0 \le s \le t$, A is a Borel subset of $\mathbb{R}$, and $N \in \mathcal{F}$ is such that $P(N) = 0$. Notice that $\mathcal{F}_s \subset \mathcal{F}_t$ if $s \le t$; that is, $(\mathcal{F}_t)_{t \ge 0}$ is a nondecreasing family of σ-fields. We say that $(\mathcal{F}_t)_{t \ge 0}$ is the *natural filtration* of Brownian motion on the probability space $(\Omega, \mathcal{F}, P)$.

Inclusion of the events of probability zero in each σ-field $\mathcal{F}_t$ has the following important consequences:

1. Any version of an adapted process is also adapted.
2. The family of σ-fields is right-continuous; that is, for all $t \ge 0$, $\cap_{s > t} \mathcal{F}_s = \mathcal{F}_t$.

Property 2 is a consequence of Blumenthal's 0–1 law (see Durrett, 2010, Theorem 8.2.3).

The natural filtration $(\mathcal{F}_t)_{t \ge 0}$ of a d-dimensional Brownian motion can be defined in a similar way.

1.6 Markov Property

Consider a Brownian motion $B = (B_t)_{t \ge 0}$. The next theorem shows that Brownian motion is an $\mathcal{F}_t$-*Markov process* with respect to its natural filtration $(\mathcal{F}_t)_{t \ge 0}$ (see Definition A.5.1).

Theorem 1.6.1 *For any measurable and bounded (or nonnegative) function $f \colon \mathbb{R} \to \mathbb{R}$, $s \ge 0$ and $t > 0$, we have*

$$E(f(B_{s+t}) | \mathcal{F}_s) = (P_t f)(B_s),$$

where

$$(P_t f)(x) = \int_{\mathbb{R}} f(y) p_t(x - y) dy.$$

Proof We have

$$E(f(B_{s+t}) | \mathcal{F}_s) = E(f(B_{s+t} - B_s + B_s) | \mathcal{F}_s).$$

Since $B_{s+t} - B_s$ is independent of $\mathcal{F}_s$, we obtain

$$E(f(B_{s+t}) | \mathcal{F}_s) = E(f(B_{s+t} - B_s + x)) |_{x = B_s}$$

$$= \int_{\mathbb{R}} f(y + B_s) \frac{1}{\sqrt{2\pi t}} e^{-|y|^2/(2t)} dy$$

$$= \int_{\mathbb{R}} f(y) \frac{1}{\sqrt{2\pi t}} e^{-|B_s - y|^2/(2t)} dy = (P_t f)(B_s),$$

which concludes the proof. $\qquad \square$

The family of operators $(P_t)_{t \geq 0}$ satisfies the semigroup property $P_t \circ P_s = P_{t+s}$ and $P_0 = \text{Id}$.

We can also show that a d-dimensional Brownian motion is an $\mathcal{F}_t$-Markov process with semigroup

$$(P_t f)(x) = \int_{\mathbb{R}^d} f(y)(2\pi t)^{-d/2} \exp\left(-\frac{|x - y|^2}{2t}\right),$$

where $f: \mathbb{R}^d \to \mathbb{R}$ is a measurable and bounded (or nonnegative) function. The transition density $p_t(x - y) = (2\pi t)^{-d/2} \exp(-|x - y|^2/(2t))$ satisfies the heat equation

$$\frac{\partial p}{\partial t} = \frac{1}{2}\Delta p, \quad t > 0,$$

with initial condition $p_0(x - y) = \delta_x(y)$.

1.7 Martingales Associated with Brownian Motion

Let $B = (B_t)_{t \geq 0}$ be a Brownian motion. The next result gives several fundamental martingales associated with Brownian motion.

Theorem 1.7.1   *The processes $(B_t)_{t \geq 0}$, $(B_t^2 - t)_{t \geq 0}$, and $(\exp(aB_t - a^2 t/2))_{t \geq 0}$, where $a \in \mathbb{R}$, are $\mathcal{F}_t$-martingales.*

Proof   Brownian motion is a martingale with respect to its natural filtration because for $s < t$

$$E(B_t - B_s|\mathcal{F}_s) = E(B_t - B_s) = 0.$$

For $B_t^2 - t$, we can write for $s < t$, using the properties of conditional expectations,

$$
\begin{aligned}
E(B_t^2|\mathcal{F}_s) &= E((B_t - B_s + B_s)^2|\mathcal{F}_s) \\
&= E((B_t - B_s)^2|\mathcal{F}_s) + 2E((B_t - B_s)B_s|\mathcal{F}_s) + E(B_s^2|\mathcal{F}_s) \\
&= E(B_t - B_s)^2 + 2B_s E((B_t - B_s)|\mathcal{F}_s) + B_s^2 \\
&= t - s + B_s^2.
\end{aligned}
$$

Finally, for $\exp(aB_t - a^2 t/2)$, we have

$$
\begin{aligned}
E(\exp(aB_t - a^2 t/2)|\mathcal{F}_s) &= e^{aB_s} E(\exp(a(B_t - B_s) - a^2 t/2)|\mathcal{F}_s) \\
&= e^{aB_s} E(\exp(a(B_t - B_s) - a^2 t/2)) \\
&= e^{aB_s} \exp(a^2(t - s)/2 - a^2 t/2) \\
&= \exp(aB_s - a^2 s/2).
\end{aligned}
$$

This concludes the proof of the theorem.   □

As an application of Theorem 1.7.1, we will study properties of the arrival time of Brownian motion at some fixed level $a \in \mathbb{R}$. This is called the Brownian *hitting time*, defined as the stopping time

$$\tau_a = \inf\{t \geq 0 : B_t = a\}.$$

The next proposition provides an explicit expression for the Laplace transform of the Brownian hitting time.

Proposition 1.7.2 *Fix $a > 0$. Then, for all $\alpha > 0$,*

$$E(\exp(-\alpha \tau_a)) = e^{-\sqrt{2\alpha}a}. \tag{1.9}$$

Proof By Theorem 1.7.1, for any $\lambda > 0$, the process $M_t = e^{\lambda B_t - \lambda^2 t/2}$ is a martingale such that

$$E(M_t) = E(M_0) = 1.$$

By the optional stopping theorem (Theorem A.7.4), for all $N \geq 1$,

$$E(M_{\tau_a \wedge N}) = 1.$$

Note that $M_{\tau_a \wedge N} = \exp(\lambda B_{\tau_a \wedge N} - \lambda^2(\tau_a \wedge N)/2) \leq e^{a\lambda}$. Moreover,

$$\lim_{N \to \infty} M_{\tau_a \wedge N} = M_{\tau_a} \quad \text{if } \tau_a < \infty,$$

$$\lim_{N \to \infty} M_{\tau_a \wedge N} = 0 \quad \text{if } \tau_a = \infty,$$

and the dominated convergence theorem implies that

$$E(\mathbf{1}_{\{\tau_a < \infty\}} M_{\tau_a}) = 1.$$

That is,

$$E\left(\mathbf{1}_{\{\tau_a < \infty\}} \exp\left(-\frac{\lambda^2 \tau_a}{2}\right)\right) = e^{-\lambda a}.$$

Letting $\lambda \downarrow 0$ we obtain

$$P(\tau_a < \infty) = 1, \tag{1.10}$$

and, consequently,

$$E\left(\exp\left(-\frac{\lambda^2 \tau_a}{2}\right)\right) = e^{-\lambda a}.$$

With the change of variable $\lambda^2/2 = \alpha$, we get the desired result. □

From expression (1.9), inverting the Laplace transform, we can compute the distribution function of the random variable τ_a:

$$P(\tau_a \le t) = \int_0^t \frac{ae^{-a^2/(2s)}}{\sqrt{2\pi s^3}} ds. \tag{1.11}$$

Furthermore, computing the derivative of (1.9) with respect to the variable α yields

$$E(\tau_a \exp(-\alpha\tau_a)) = \frac{ae^{-\sqrt{2\alpha}a}}{\sqrt{2\alpha}},$$

and letting $\alpha \downarrow 0$ we obtain $E(\tau_a) = +\infty$.

Proposition 1.7.3 *If $a < 0 < b$ then*

$$P(\tau_a < \tau_b) = \frac{b}{b-a}.$$

Proof By (1.10) we have that $\tau_b < \infty$ almost surely, and taking into account that $(-B_t)_{t\ge0}$ is also a Brownian motion, we deduce that $\tau_a < \infty$ almost surely. By the optional stopping theorem (Theorem A.7.4), we have

$$E(B_{t\wedge\tau_a\wedge\tau_b}) = E(B_0) = 0.$$

Since, for all $t \ge 0$,

$$a \le B_{t\wedge\tau_a\wedge\tau_b} \le b,$$

letting $t \to \infty$ and using the dominated convergence theorem, it follows that

$$E(B_{\tau_a\wedge\tau_b}) = 0.$$

This implies that

$$0 = aP(\tau_a < \tau_b) + b(1 - P(\tau_a < \tau_b)),$$

which proves the desired formula. $\qquad\square$

Proposition 1.7.4 *Let $T = \inf\{t \ge 0 : B_t \notin (a, b)\}$, where $a < 0 < b$. Then*

$$E(T) = -ab.$$

Proof Because $B_t^2 - t$ is a martingale, we get, by the optional stopping theorem (Theorem A.7.4),

$$E(B_{T\wedge t}^2) = E(T \wedge t).$$

Therefore, from the dominated convergence theorem and Proposition 1.7.3,

$$E(T) = \lim_{t\to\infty} E(B_{T\wedge t}^2) = E(B_T^2) = -ab,$$

which concludes the proof.   □

1.8 Strong Markov Property

Let $B = (B_t)_{t \geq 0}$ be a Brownian motion. The next result is the *strong Markov property* of Brownian motion, which was first proved independently by Hunt (1956) and Dynkin and Yushkevich (1956).

Theorem 1.8.1 *Let T be a finite stopping time with respect to the natural filtration of Brownian motion $(\mathcal{F}_t)_{t \geq 0}$. Then the process*

$$B_{T+t} - B_T, \quad t \geq 0,$$

is a Brownian motion that is independent of $\mathcal{F}_T$.

Proof   Consider the process $\tilde{B} = (\tilde{B}_t)_{t \geq 0}$ defined by $\tilde{B}_t = B_{T+t} - B_T$, and suppose first that T is bounded. Let $\lambda \in \mathbb{R}$ and $0 \leq s \leq t$. Applying the optional stopping theorem to the complex-valued martingale

$$\exp\left(i\lambda \tilde{B}_t + \frac{\lambda^2 t}{2}\right)$$

yields

$$\mathrm{E}\left(\exp\left(i\lambda B_{T+t} + \frac{\lambda^2}{2}(T+t)\right)\Big|\mathcal{F}_{T+s}\right) = \exp\left(i\lambda B_{T+s} + \frac{\lambda^2}{2}(T+s)\right).$$

Therefore,

$$\mathrm{E}\left(\exp\left(i\lambda(B_{T+t} - B_{T+s})\right)\Big|\mathcal{F}_{T+s}\right) = \exp\left(-\frac{\lambda^2}{2}(t-s)\right).$$

This implies that the increments of $\tilde{B}$ are independent, stationary, and normally distributed with mean zero and variance equal to the length of the increment. Moreover the process $\tilde{B}$ is independent of $\mathcal{F}_T$, which concludes the proof when T is bounded. If T is not bounded, we can consider the stopping time $T \wedge N$ and let $N \to \infty$.   □

As a consequence, for any measurable and bounded (or nonnegative) function $f : \mathbb{R} \to \mathbb{R}$ and any finite stopping time T for the filtration $(\mathcal{F}_t)_{t \geq 0}$, we have

$$\mathrm{E}(f(B_{T+t})|\mathcal{F}_T) = (P_t f)(B_T),$$

where P_t is the semigroup of operators associated with Brownian motion.

As an application of the strong Markov property, we have the following *reflection principle*, which was first formulated by Lévy (1939).

Theorem 1.8.2 *Let $M_t = \sup_{0 \leq s \leq t} B_s$. Then, for all $a > 0$,*

$$P(M_t \geq a) = 2P(B_t > a).$$

Proof Consider the reflected process

$$\hat{B}_t = B_t \mathbf{1}_{\{t \leq \tau_a\}} + (2a - B_t)\mathbf{1}_{\{t > \tau_a\}}, \quad t \geq 0.$$

Recall that $\tau_a < \infty$ a.s. by (1.10). Then, by the strong Markov property (Theorem 1.8.1), both the processes $(B_{t+\tau_a} - a)_{t \geq 0}$ and $(-B_{t+\tau_a} + a)_{t \geq 0}$ are Brownian motions that are independent of B_{τ_a}. Pasting the first process to the end point of $(B_t)_{t \in [0,\tau_a]}$, and doing the same with the second process, yields two processes with the same distribution. The first is just $(B_t)_{t \geq 0}$ and the second is $\hat{B} = (\hat{B}_t)_{t \geq 0}$. Thus, we conclude that $\hat{B}$ is also a Brownian motion. Therefore,

$$P(M_t \geq a) = P(B_t > a) + P(M_t \geq a, B_t \leq a)$$
$$= P(B_t > a) + P(\hat{B}_t \geq a) = 2P(B_t > a),$$

which concludes the proof. □

Corollary 1.8.3 *For any $a > 0$, the random variable $M_a = \sup_{t \in [0,a]} B_t$ has density*

$$p(x) = \frac{2}{\sqrt{2\pi a}} e^{-x^2/(2a)} \mathbf{1}_{[0,\infty)}(x).$$

Using the reflection principle (Theorem 1.8.2), we obtain the following property.

Lemma 1.8.4 *With probability one, Brownian motion attains its maximum on $[0, 1]$ at a unique point.*

Proof It suffices to show that the set

$$G = \left\{ \omega : \sup_{t \in [0,1]} B_t = B_{t_1} = B_{t_2} \text{ for some } t_1 \neq t_2 \right\}$$

has probability zero. For each $n \geq 0$, we denote by $\mathcal{I}_n$ the set of dyadic intervals of the form $[(j-1)2^{-n}, j2^{-n}]$, with $1 \leq j \leq 2^n$. The set G is equal to the countable union

$$\bigcup_{n \geq 1} \bigcup_{I_1, I_2 \in \mathcal{I}_n, I_1 \cap I_2 = \emptyset} \left\{ \sup_{t \in I_1} B_t = \sup_{t \in I_2} B_t \right\}.$$

Therefore, it suffices to check that, for each $n \geq 1$ and for any pair of disjoint intervals I_1, I_2,

$$P\left(\sup_{t \in I_1} B_t = \sup_{t \in I_2} B_t \right) = 0. \tag{1.12}$$

Property (1.12) is a consequence of the fact that, for any rectangle $[a, b] \subset [0, 1]$, the law of the random variable $\sup_{t \in [a,b]} B_t$ conditioned on $\mathcal{F}_a$ is continuous. To establish this property, it suffices to write

$$\sup_{t \in [a,b]} B_t = \sup_{t \in [a,b]} (B_t - B_a) + B_a.$$

Then, conditioning on $\mathcal{F}_a$, B_a is a constant and $\sup_{t \in [a,b]}(B_t - B_a)$ has the same law as $\sup_{0 \leq t \leq b-a} B_t$, which has the density given in Corollary 1.8.3.

□

Exercises

1.1 Let Z be a random variable with law $N(0, 1)$. Consider the process $X_t = \sqrt{t}Z$, $t \geq 0$. Is $(X_t)_{t \geq 0}$ a Brownian motion? Which properties of Definition 1.2.1 hold and which don't?

1.2 Let B be a d-dimensional Brownian motion. Consider an orthogonal $d \times d$ matrix U (that is, $UU^T = I_d$, where I_d denotes the identity matrix of order d). Show that the process

$$X_t = UB_t, \quad t \geq 0,$$

is a d-dimensional Brownian motion.

1.3 Compute the mean and covariance function of the following stochastic processes related to Brownian motion.

(a) *Brownian bridge*: $X_t = B_t - tB_1, t \in [0, 1]$.

(b) *Brownian motion with drift*: $X_t = \sigma B_t + \mu t, t \geq 0$, where $\sigma > 0$ and $\mu \in \mathbb{R}$ are constants.

(c) *Geometric Brownian motion*: $X_t = e^{\sigma B_t + \mu t}, t \geq 0$, where $\sigma > 0$ and $\mu \in \mathbb{R}$ are constants.

Which of the above are Gaussian processes?

1.4 Define $X_t = \int_0^t B_s ds$, where $B = (B_t)_{t \geq 0}$ is a Brownian motion. Show that X_t is a Gaussian random variable. Compute its mean and its variance.

1.5 Let B be a Brownian motion. Show that, for any $a > 0$, the process $(a^{-1/2} B_{at})_{t \geq 0}$ is a Brownian motion.

1.6 Let B be a Brownian motion. Show that, for any $h > 0$, the process $(B_{t+h} - B_h)_{t \geq 0}$ is a Brownian motion.

1.7 Let B be a Brownian motion. Show that the process $(-B_t)_{t \geq 0}$ is a Brownian motion.

1.8 Let B be a Brownian motion. Show that $\lim_{t \to \infty} B_t/t = 0$ almost surely and that the process

$$X_t = \begin{cases} tB_{1/t} & \text{if } t > 0, \\ 0 & \text{if } t = 0, \end{cases}$$

is a Brownian motion.

1.9 Let B be a Brownian motion. Show that

$$P\left(\sup_{s,t\in[0,1]} \frac{|B_t - B_s|}{\sqrt{|t - s|}} = +\infty\right) = 1.$$

1.10 Let B be a Brownian motion. Using the Borel–Cantelli lemma, show that if $(\pi^n)_{n\geq 1}$, $\pi^n = \{0 = t_0^n < \cdots < t_{k_n}^n = t\}$, is a sequence of partitions of $[0, t]$ such that $\sum_n |\pi^n| < \infty$, then $\sum_{j=0}^{k_n-1}(B_{t_{j+1}^n} - B_{t_j^n})^2$ converges almost surely to t.

1.11 Let $(\Omega, \mathcal{F}, P)$ be the Wiener space. Show that $\mathcal{F}$ coincides with the σ-field generated by the collection of cylinder sets

$$\{\omega \in \Omega : \omega(t_1) \in A_1, \ldots, \omega(t_k) \in A_k\}$$

where $k \geq 1$ is an integer, $A_1, \ldots, A_k \in \mathcal{B}(\mathbb{R})$, and $0 \leq t_1 < \cdots < t_k$.

2

Stochastic Calculus

The first aim of this chapter is to construct Itô's stochastic integrals of the form $\int_0^\infty u_t dB_t$, where $B = (B_t)_{t\geq 0}$ is a Brownian motion and $u = (u_t)_{t\geq 0}$ is an adapted process. We then prove Itô's formula, which is a change of variables formula for Itô's stochastic integrals and plays a crucial role in the applications of stochastic calculus. We then discuss some consequences of Itô's formula, including Tanaka's formula, the Stratonovich integral, and the integral representation theorem for square integrable random variables. Finally, we present Girsanov's theorem, which provides a change of probability under which a Brownian motion with drift becomes a Brownian motion.

2.1 Stochastic Integrals

The stochastic integral of adapted processes with respect to Brownian motion originates in Itô (1944). For complete expositions of this topic we refer to Ikeda and Watanabe (1989), Karatzas and Shreve (1998), and Baudoin (2014).

Recall that $B = (B_t)_{t\geq 0}$ is a Brownian motion defined on a probability space $(\Omega, \mathcal{F}, P)$ equipped with its natural filtration $(\mathcal{F}_t)_{t\geq 0}$. We proved in Chapter 1 that the trajectories of Brownian motion have infinite variation on any finite interval. So, in general, we cannot define the integral

$$\int_0^T u_t(\omega) dB_t(\omega)$$

as a pathwise integral. However, we will construct the integral $\int_0^\infty u_t dB_t$ by means of a global probabilistic approach for a class of processes satisfying some adaptability and integrability conditions specified below.

Definition 2.1.1 We say that a stochastic process $u = (u_t)_{t\geq 0}$ is *progres-*

sively measurable if, for any $t \geq 0$, the restriction of u to $\Omega \times [0, t]$ is $\mathcal{F}_t \times \mathcal{B}([0, t])$-measurable.

Remark 2.1.2 If u is adapted and measurable (i.e., the mapping $(\omega, s) \longrightarrow u_s(\omega)$ is measurable on the product space $\Omega \times \mathbb{R}_+$ with respect to the product σ-field $\mathcal{F} \times \mathcal{B}(\mathbb{R}_+)$) then there is a version of u which is progressively measurable (see Meyer, 1984, Theorem 4.6). Progressive measurability guarantees that random variables of the form $\int_0^t u_s ds$ are $\mathcal{F}_t$-measurable.

Let $\mathcal{P}$ be the σ-field of sets $A \subset \Omega \times \mathbb{R}_+$ such that $\mathbf{1}_A$ is progressively measurable. We denote by $L^2(\mathcal{P})$ the Hilbert space $L^2(\Omega \times \mathbb{R}_+, \mathcal{P}, P \times \ell)$, where ℓ is the Lebesgue measure, equipped with the norm

$$\|u\|^2_{L^2(\mathcal{P})} = E\left(\int_0^\infty u_s^2 ds\right) = \int_0^\infty E(u_s^2) ds,$$

where the last equality follows from Fubini's theorem.

In this section we define the stochastic integral $\int_0^\infty u_t dB_t$ of a process u in $L^2(\mathcal{P})$ as the limit in $L^2(\Omega)$ of integrals of simple processes.

Definition 2.1.3 A process $u = (u_t)_{t \geq 0}$ is called a *simple process* if it is of the form

$$u_t = \sum_{j=0}^{n-1} \phi_j \mathbf{1}_{(t_j, t_{j+1}]}(t), \tag{2.1}$$

where $0 \leq t_0 < t_1 < \cdots < t_n$, and the ϕ_j are $\mathcal{F}_{t_j}$-measurable random variables such that $E(\phi_j^2) < \infty$. We denote by $\mathcal{E}$ the space of simple processes.

We define the stochastic integral of a process $u \in \mathcal{E}$ of the form (2.1) as

$$I(u) := \int_0^\infty u_t dB_t = \sum_{j=0}^{n-1} \phi_j (B_{t_{j+1}} - B_{t_j}).$$

The stochastic integral defined on the space $\mathcal{E}$ of simple processes has the following three properties.

1 Linearity

For any $a, b \in \mathbb{R}$ and simple processes $u, v \in \mathcal{E}$,

$$\int_0^\infty (au_t + bv_t) \, dB_t = a \int_0^\infty u_t dB_t + b \int_0^\infty v_t dB_t.$$

2 Zero Mean

For any $u \in \mathcal{E}$,

$$\mathrm{E}\left(\int_0^\infty u_t dB_t \right) = 0.$$

In fact, assuming that u is given by (2.1), and taking into account that the random variables ϕ_j and $B_{t_{j+1}} - B_{t_j}$ are independent, we obtain

$$\mathrm{E}\left(\int_0^\infty u_t dB_t \right) = \sum_{j=0}^{n-1} \mathrm{E}(\phi_j(B_{t_{j+1}} - B_{t_j})) = \sum_{j=0}^{n-1} \mathrm{E}(\phi_j)\mathrm{E}(B_{t_{j+1}} - B_{t_j}) = 0.$$

3 Isometry Property

For any $u \in \mathcal{E}$,

$$\mathrm{E}\left(\left(\int_0^\infty u_t dB_t \right)^2 \right) = \mathrm{E}\left(\int_0^\infty u_t^2 dt \right).$$

Proof   Assume that u is given by (2.1). Set $\Delta B_j = B_{t_{j+1}} - B_{t_j}$. Then

$$\mathrm{E}\left(\phi_i \phi_j \Delta B_i \Delta B_j \right) = \begin{cases} 0 & \text{if } i \neq j, \\ \mathrm{E}(\phi_j^2)(t_{j+1} - t_j) & \text{if } i = j, \end{cases}$$

because if $i < j$ the random variables $\phi_i \phi_j \Delta B_i$ and ΔB_j are independent, and if $i = j$ the random variables ϕ_i^2 and $(\Delta B_i)^2$ are independent. So, we obtain

$$\mathrm{E}\left(\left(\int_0^\infty u_t dB_t \right)^2 \right) = \sum_{i,j=0}^{n-1} \mathrm{E}\left(\phi_i \phi_j \Delta B_i \Delta B_j \right) = \sum_{i=0}^{n-1} \mathrm{E}(\phi_i^2)(t_{i+1} - t_i)$$

$$= \mathrm{E}\left(\int_0^\infty u_t^2 dt \right),$$

which concludes the proof of the isometry property.   □

The extension of the stochastic integral to the class $L^2(\mathcal{P})$ is based on the following density result.

Proposition 2.1.4   *The space $\mathcal{E}$ of simple processes is dense in $L^2(\mathcal{P})$.*

Proof   We first establish that any $u \in L^2(\mathcal{P})$ can be approximated by processes which are continuous in $L^2(\Omega)$. Then our result will follow if we show that simple processes are dense in the space of processes which are continuous in $L^2(\Omega)$.

If u belongs to $L^2(\mathcal{P})$, we define

$$u_t^{(n)} = n \int_{(t-1/n)\vee 0}^t u_s \, ds.$$

The processes $u_t^{(n)}$ are continuous in $L^2(\Omega)$ and satisfy

$$\lim_{n\to\infty} \mathrm{E}\left(\int_0^\infty |u_t - u_t^{(n)}|^2 dt \right) = 0.$$

Indeed, for each ω we have

$$\int_0^\infty |u_t(\omega) - u_t^{(n)}(\omega)|^2 dt \xrightarrow{n\to\infty} 0,$$

and we can apply the dominated convergence theorem because

$$\int_0^\infty |u_t^{(n)}(\omega)|^2 dt \le \int_0^\infty |u_t(\omega)|^2 dt.$$

Suppose that $u \in L^2(\mathcal{P})$ is continuous in $L^2(\Omega)$. In this case, we can choose approximating processes $u_t^{(n,N)} \in \mathcal{E}$ defined by

$$u_t^{(n,N)} = \sum_{j=0}^{n-1} u_{t_j} \mathbf{1}_{(t_j, t_{j+1}]}(t),$$

where $t_j = jN/n$. The continuity in $L^2(\Omega)$ of u implies that

$$\mathrm{E}\left(\int_0^\infty |u_t - u_t^{(n,N)}|^2 \, dt \right) \le \mathrm{E}\left(\int_N^\infty u_t^2 dt \right) + N \sup_{|t-s|\le N/n} \mathrm{E}(|u_t - u_s|^2).$$

This converges to zero if we first let $n \to \infty$ and then $N \to \infty$. $\square$

Proposition 2.1.5 *The stochastic integral can be extended to a linear isometry*

$$I: L^2(\mathcal{P}) \to L^2(\Omega).$$

Proof The stochastic integral of a process u in $L^2(\mathcal{P})$ is defined as the following limit in $L^2(\Omega)$:

$$I(u) = \lim_{n\to\infty} \int_0^\infty u_t^{(n)} dB_t, \tag{2.2}$$

where $u^{(n)}$ is a sequence of simple processes which converges to u in the norm of $L^2(\mathcal{P})$. Notice that the limit (2.2) exists because the sequence of random variables $\int_0^\infty u_t^{(n)} dB_t$ is Cauchy in $L^2(\Omega)$, owing to the isometry property

$$E\left(\left(\int_0^\infty u_t^{(n)} dB_t - \int_0^\infty u_t^{(m)} dB_t\right)^2\right) = E\left(\int_0^\infty (u_t^{(n)} - u_t^{(m)})^2 dt\right)$$

$$\leq 2E\left(\int_0^\infty (u_t^{(n)} - u_t)^2 dt\right) + 2E\left(\int_0^\infty (u_t - u_t^{(m)})^2 dt\right) \overset{n,m \to \infty}{\longrightarrow} 0.$$

Furthermore, it is easy to show that the limit (2.2) does not depend on the approximating sequence $u^{(n)}$. $\square$

The stochastic integral has the following properties: for any $u, v \in L^2(\mathcal{P})$,

$$E(I(u)) = 0 \quad \text{and} \quad E(I(u)I(v)) = E\left(\int_0^\infty u_s v_s ds\right).$$

For any $T > 0$, we set

$$\int_0^T u_s dB_s = \int_0^\infty u_s \mathbf{1}_{[0,T]}(s) dB_s, \tag{2.3}$$

which is the indefinite integral of u with respect to B. Notice that in order to define $\int_0^T u_s dB_s$, we only need that $u \in L_T^2(\mathcal{P})$, where

$$L_T^2(\mathcal{P}) := L^2(\Omega \times [0, T], \mathcal{P}|_{\Omega \times [0,T]}, P \times \ell).$$

Example 2.1.6 For any $T > 0$, we have

$$\int_0^T B_t dB_t = \tfrac{1}{2} B_T^2 - \tfrac{1}{2} T.$$

Indeed, the process B being continuous in $L^2(\Omega)$, we can choose as approximating sequence

$$u_t^{(n)} = \sum_{j=0}^{n-1} B_{t_j} \mathbf{1}_{(t_j, t_{j+1}]}(t),$$

where $t_j = jT/n$, and we obtain, using Proposition 1.2.4,

$$\int_0^T B_t dB_t = \lim_{n \to \infty} \sum_{j=0}^{n-1} B_{t_j}(B_{t_{j+1}} - B_{t_j})$$

$$= \tfrac{1}{2} \lim_{n \to \infty} \sum_{j=0}^{n-1} (B_{t_{j+1}}^2 - B_{t_j}^2) - \tfrac{1}{2} \lim_{n \to \infty} \sum_{j=0}^{n-1} (B_{t_{j+1}} - B_{t_j})^2$$

$$= \tfrac{1}{2} B_T^2 - \tfrac{1}{2} T,$$

where the convergence holds in $L^2(\Omega)$.

2.2 Indefinite Stochastic Integrals

We denote by $L_\infty^2(\mathcal{P})$ the set of progressively measurable processes such that $E\left(\int_0^t u_s^2 ds\right) < \infty$ for each $t > 0$. For any process $u \in L_\infty^2(\mathcal{P})$, we can define the indefinite integral process

$$\left\{\int_0^t u_s dB_s, t \geq 0\right\}.$$

We now give six properties of indefinite integrals.

1 Additivity

For any $a \leq b \leq c$, we have

$$\int_a^b u_s dB_s + \int_b^c u_s dB_s = \int_a^c u_s dB_s.$$

2 Factorization

If $a < b$, and F is a bounded and $\mathcal{F}_a$-measurable random variable then

$$\int_a^b F u_s dB_s = F \int_a^b u_s dB_s.$$

3 Martingale Property

Proposition 2.2.1 *Let $u \in L_\infty^2(\mathcal{P})$. Then, the indefinite stochastic integral*

$$M_t = \int_0^t u_s dB_s, \quad t \geq 0,$$

is a square integrable martingale with respect to the filtration $(\mathcal{F}_t)_{t \geq 0}$ and admits a continuous version.

Proof We first prove the martingale property. Suppose that $u \in \mathcal{E}$ is a simple process of the form

$$u_t = \sum_{j=0}^{n-1} \phi_j \mathbf{1}_{(t_j, t_{j+1}]}(t).$$

Then, for any $s \leq t$,

$$\mathrm{E}\left(\int_0^t u_v dB_v \Big| \mathcal{F}_s\right) = \sum_{j=0}^{n-1} \mathrm{E}\left(\phi_j(B_{t_{j+1}\wedge t} - B_{t_j\wedge t})\big|\mathcal{F}_s\right)$$

$$= \sum_{j=0}^{n-1} \mathrm{E}\left(\mathrm{E}\left(\phi_j(B_{t_{j+1}\wedge t} - B_{t_j\wedge t})\big|\mathcal{F}_{t_j\vee s}\right)\big|\mathcal{F}_s\right)$$

$$= \sum_{j=0}^{n-1} \mathrm{E}\left(\phi_j\mathrm{E}\left(B_{t_{j+1}\wedge t} - B_{t_j\wedge t}\big|\mathcal{F}_{t_j\vee s}\right)\big|\mathcal{F}_s\right)$$

$$= \sum_{j=0}^{n-1} \phi_j(B_{t_{j+1}\wedge s} - B_{t_j\wedge s}) = \int_0^s u_v dB_v.$$

So, $M_t = \int_0^t u_s dB_s$ is an $\mathcal{F}_t$-martingale if $u \in \mathcal{E}$.

Fix $T > 0$. In the general case, let $u^{(n)}$ be a sequence of simple processes that converges to u in $L_T^2(\mathcal{P})$. Then, for any $t \in [0, T]$,

$$\int_0^t u_s^{(n)} dB_s \xrightarrow{L^2(\Omega)} \int_0^t u_s dB_s.$$

Taking into account that the above convergence in $L^2(\Omega)$ implies the convergence in $L^2(\Omega)$ of the conditional expectations, we deduce that the process $\left(\int_0^t u_s dB_s\right)_{t\in[0,T]}$ is a martingale. This holds for any $T > 0$, which implies that the indefinite integral process is a martingale on $\mathbb{R}_+$.

Let us prove that the indefinite integral has a continuous version. Let $u \in L_\infty^2(\mathcal{P})$ and fix $T > 0$. Consider a sequence of simple processes $u^{(n)}$ which converges to u in $L_T^2(\mathcal{P})$. By the continuity of the paths of Brownian motion, the stochastic integral $M_t^{(n)} = \int_0^t u_s^{(n)} dB_s$ has continuous trajectories. Then, taking into account that $M^{(n)}$ is a martingale, Doob's maximal inequality (Theorem A.7.5) yields, for any $\lambda > 0$,

$$\mathrm{P}\left(\sup_{0\leq t\leq T} |M_t^{(n)} - M_t^{(m)}| > \lambda\right) \leq \frac{1}{\lambda^2}\mathrm{E}\left(|M_T^{(n)} - M_T^{(m)}|^2\right)$$

$$= \frac{1}{\lambda^2}\mathrm{E}\left(\int_0^T |u_t^{(n)} - u_t^{(m)}|^2 dt\right) \xrightarrow{n,m\to\infty} 0.$$

We can choose an increasing sequence of natural numbers $n_k, k = 1, 2, \ldots,$ such that

$$\mathrm{P}\left(\sup_{0\leq t\leq T} |M_t^{(n_{k+1})} - M_t^{(n_k)}| > 2^{-k}\right) \leq 2^{-k}.$$

The events $A_k := \left\{ \sup_{0 \leq t \leq T} |M_t^{(n_{k+1})} - M_t^{(n_k)}| > 2^{-k} \right\}$ verify that

$$\sum_{k=1}^{\infty} P(A_k) < \infty.$$

Hence, the Borel–Cantelli lemma implies that $P(\limsup_{k \to \infty} A_k) = 0$. Set $N = \limsup_{k \to \infty} A_k$. Then, for any $\omega \notin N$, there exists $k_1(\omega)$ such that, for all $k \geq k_1(\omega)$,

$$\sup_{0 \leq t \leq T} |M_t^{(n_{k+1})}(\omega) - M_t^{(n_k)}(\omega)| \leq 2^{-k}.$$

As a consequence, if $\omega \notin N$, the sequence $M_t^{(n_k)}(\omega)$ is uniformly convergent on $[0, T]$ to a continuous function $J_t(\omega)$. Moreover, we know that, for any $t \in [0, T]$, $M_t^{(n_k)}$ converges in $L^2(\Omega)$ to $\int_0^t u_s dB_s$. So $J_t(\omega) = \int_0^t u_s dB_s$ almost surely, for all $t \in [0, T]$. Since $T > 0$ is arbitrary, this implies the existence of a continuous version for $(M_t)_{t \geq 0}$. □

4 Maximal Inequalities

For any $T, \lambda > 0$, and $u \in L^2_{\infty}(\mathcal{P})$,

$$P\left(\sup_{t \in [0,T]} |M_t| > \lambda \right) \leq \frac{1}{\lambda^2} E\left(\int_0^T u_t^2 dt \right) \tag{2.4}$$

and

$$E\left(\sup_{t \in [0,T]} |M_t|^2 \right) \leq 4E\left(\int_0^T u_t^2 dt \right). \tag{2.5}$$

These inequalities are a direct consequence of Proposition 2.2.1 and Doob's maximal inequalities (Theorem A.7.5). We remark that if u belongs to $L^2(\mathcal{P})$ then these inequalities also hold if T is replaced by ∞.

5 Quadratic Variation of the Integral Process

Proposition 2.2.2 *Let $u \in L^2_{\infty}(\mathcal{P})$. Consider the following subdivision of the interval $[0, t]$:*

$$\pi = \{0 = t_0 < t_1 < \cdots < t_n = t\}.$$

Then, as $|\pi| \to 0$,

$$S_{\pi}^2(u) := \sum_{j=0}^{n-1} \left(\int_{t_j}^{t_{j+1}} u_s dB_s \right)^2 \xrightarrow{L^1(\Omega)} \int_0^t u_s^2 ds. \tag{2.6}$$

This proposition implies that the quadratic variation of the martingale M_t equals $\int_0^t u_s^2 ds$ (see Theorem A.7.2).

Proof of Proposition 2.2.2 As a consequence of the quadratic variation property of Brownian motion (see Proposition 1.2.4), it is easy to see that (2.6) holds when $u \in \mathcal{E}$ (see Exercise 2.2). Let $u \in L^2_\infty(\mathcal{P})$. Fix $\epsilon > 0$ and let $v \in \mathcal{E}$ such that

$$\mathrm{E}\left(\int_0^t (u_s - v_s)^2 ds\right) < \epsilon. \tag{2.7}$$

Then, we write

$$\mathrm{E}\left(\left|S_\pi^2(u) - \int_0^t u_s^2 ds\right|\right) \le \mathrm{E}\left(\left|S_\pi^2(u) - S_\pi^2(v)\right|\right) + \mathrm{E}\left(\left|S_\pi^2(v) - \int_0^t v_s^2 ds\right|\right)$$
$$+ \mathrm{E}\left(\left|\int_0^t (u_s^2 - v_s^2) ds\right|\right).$$

As (2.6) holds for $v \in \mathcal{E}$, the second term converges to 0 as $|\pi| \to 0$. Moreover, since $u_s^2 - v_s^2 = (u_s + v_s)(u_s - v_s)$, using the Cauchy–Schwarz inequality and (2.7), we obtain

$$\mathrm{E}\left(\left|\int_0^t (u_s^2 - v_s^2) ds\right|\right) \le \sqrt{\epsilon}\left(\mathrm{E}\left(\int_0^t (u_s + v_s)^2 ds\right)\right)^{1/2}$$
$$\le \sqrt{\epsilon}\left(\left(\mathrm{E}\left(\int_0^t u_s^2 ds\right)\right)^{1/2} + \sqrt{\epsilon}\right).$$

Similarly,

$$\mathrm{E}\left(\left|S_\pi^2(u) - S_\pi^2(v)\right|\right) \le \sqrt{\epsilon}\left(\left(\mathrm{E}\left(\int_0^t u_s^2 ds\right)\right)^{1/2} + \sqrt{\epsilon}\right).$$

Therefore,

$$\limsup_{|\pi| \to 0} \mathrm{E}\left(\left|S_\pi^2(u) - \int_0^t u_s^2 ds\right|\right) \le 2\sqrt{\epsilon}\left(\left(\mathrm{E}\left(\int_0^t u_s^2 ds\right)\right)^{1/2} + \sqrt{\epsilon}\right).$$

Taking into account that $\epsilon > 0$ is arbitrary, we get the desired convergence.
□

As a consequence of this proposition and the Burkholder–Davis–Gundy inequality (Theorem A.7.6), we obtain the following.

Theorem 2.2.3 *Let $u \in L^2_\infty(\mathcal{P})$. Then, for any $p > 0$ and $T > 0$, we have*

$$c_p \mathrm{E}\left(\left|\int_0^T u_s^2 ds\right|^{p/2}\right) \le \mathrm{E}\left(\sup_{t \in [0,T]} \left|\int_0^t u_s dB_s\right|^p\right) \le C_p \mathrm{E}\left(\left|\int_0^T u_s^2 ds\right|^{p/2}\right).$$

6 Stochastic Integration up to a Stopping Time

Proposition 2.2.4 *Suppose that $u \in L^2(\mathcal{P})$ and let τ be a finite stopping time. Then the process $u\mathbf{1}_{[0,\tau]}$ also belongs to $L^2(\mathcal{P})$ and we have*

$$\int_0^\infty u_t \mathbf{1}_{[0,\tau]}(t) dB_t = \int_0^\tau u_t dB_t.$$

Proof Suppose first that $u_t = F\mathbf{1}_{(a,b]}(t)$, where $0 \le a < b$ and $F \in L^2(\Omega, \mathcal{F}_a, P)$, and τ takes values in a finite set $\{0 = t_0 < t_1 < \cdots < t_n\}$. On the one hand, we have

$$\int_0^\tau u_t dB_t = F(B_{b\wedge\tau} - B_{a\wedge\tau}).$$

On the other hand, the process $\mathbf{1}_{(0,\tau]}$ is simple because

$$\mathbf{1}_{(0,\tau]}(t) = \sum_{j=0}^{n-1} \mathbf{1}_{\{\tau \ge t_{j+1}\}} \mathbf{1}_{(t_j,t_{j+1}]}(t)$$

and $\mathbf{1}_{\{\tau \ge t_{j+1}\}} = \mathbf{1}_{\{\tau \le t_j\}^c} \in \mathcal{F}_{t_j}$. Therefore,

$$\int_0^\infty u_t \mathbf{1}_{(0,\tau]}(t) dB_t = F \sum_{j=0}^{n-1} \mathbf{1}_{\{\tau \ge t_{j+1}\}} \int_0^\infty \mathbf{1}_{(a,b] \cap (t_j,t_{j+1}]}(t) dB_t$$

$$= F \sum_{i=1}^{n} \mathbf{1}_{\{\tau = t_i\}} \int_0^\infty \mathbf{1}_{(a,b] \cap [0,t_i]}(t) dB_t$$

$$= F(B_{b\wedge\tau} - B_{a\wedge\tau}).$$

For a general finite stopping time τ, we approximate τ by the sequence of stopping times

$$\tau_n = \sum_{i=1}^{n2^n} \frac{i}{2^n} \mathbf{1}_{\{(i-1)/2^n \le \tau < i/2^n\}},$$

that satisfy $\tau_n \downarrow \tau$. Taking the limit as n tends to infinity we deduce the equality in the case of a simple process.

In the case of a general process $u \in L^2(\mathcal{P})$, we approximate u by simple processes $u^{(n)}$ in the norm of $L^2(\mathcal{P})$. The convergence

$$\int_0^\tau u_t^{(n)} dB_t \overset{L^2(\Omega)}{\longrightarrow} \int_0^\tau u_t dB_t$$

follows from the maximal inequality (2.5) when $T = \infty$. In fact,

$$\mathrm{E}\left(\left|\int_0^\tau u_t^{(n)} dB_t - \int_0^\tau u_t dB_t\right|^2\right) \leq \mathrm{E}\left(\sup_{t \geq 0}\left|\int_0^t (u_s^{(n)} - u_s) dB_s\right|^2\right)$$

$$\leq 4\mathrm{E}\left(\int_0^\infty |u_s^{(n)} - u_s|^2 ds\right),$$

which concludes the proof.   □

We observe that Proposition 2.2.4 also holds when $u \in L_\infty^2(\mathcal{P})$ and τ is bounded.

2.3 Integral of General Processes

The stochastic integral can be defined for a class of processes larger than $L_\infty^2(\mathcal{P})$. Let $L_{loc}^2(\mathcal{P})$ be the set of progressively measurable processes $u = (u_t)_{t \geq 0}$ such that for all $t \geq 0$,

$$\mathrm{P}\left(\int_0^t u_s^2 ds < \infty\right) = 1.$$

Suppose that $u \in L_{loc}^2(\mathcal{P})$. For each $n \geq 1$, we define the stopping time

$$T_n = \inf\left\{t \geq 0 : \int_0^t u_s^2 ds = n\right\} \tag{2.8}$$

and the sequence of processes $u_t^{(n)} = u_t \mathbf{1}_{[0,T_n]}(t)$ which belong to $L^2(\mathcal{P})$.

Proposition 2.3.1 *Let $u \in L_{loc}^2(\mathcal{P})$. Then there exists an adapted and continuous process $\left(\int_0^t u_s dB_s\right)_{t \geq 0}$ such that, for any $n \geq 1$,*

$$\int_0^t u_s^{(n)} dB_s = \int_0^t u_s dB_s \quad on \ t \leq T_n.$$

Proof If $n \leq m$, as a consequence of Proposition 2.2.4 we have that, on the set $\{t \leq T_n\}$,

$$\int_0^t u_s^{(n)} dB_s = \int_0^t u_s^{(m)} dB_s. \tag{2.9}$$

We define

$$\int_0^t u_s dB_s = \int_0^t u_s^{(n)} dB_s \quad on \ t \leq T_n.$$

By (2.9), this definition does not depend on n and produces an adapted and continuous process because $T_n \uparrow \infty$.   □

If $u \in L^2_{loc}(\mathcal{P})$, the process $(M_t)_{t\geq0}$ defined by $M_t = \int_0^t u_s dB_s$ is a continuous local martingale; that is, there exists a sequence of stopping times $T_n \uparrow \infty$ (we can take the stopping times defined in (2.8)), such that, for each $n \geq 1$, $(M_{t \wedge T_n})_{t\geq0}$ is a martingale.

Instead of the isometry property, the stochastic integral of processes in $L^2_{loc}(\mathcal{P})$ has the following continuity property in probability.

Proposition 2.3.2 *Suppose that $u \in L^2_{loc}(\mathcal{P})$. Then, for all $K, \delta, T > 0$, we have*

$$P\left(\left|\int_0^T u_s dB_s\right| \geq K\right) \leq P\left(\int_0^T u_s^2 ds \geq \delta\right) + \frac{\delta}{K^2}.$$

Proof Consider the stopping time defined by

$$\tau = \inf\left\{t \geq 0 : \int_0^t u_s^2 ds = \delta\right\},$$

with the convention that $\tau = T$ if $\int_0^T u_s^2 ds < \delta$. We have on the one hand

$$P\left(\left|\int_0^T u_s dB_s\right| \geq K\right) \leq P\left(\int_0^T u_s^2 ds \geq \delta\right)$$
$$+ P\left(\left|\int_0^T u_s dB_s\right| \geq K, \int_0^T u_s^2 ds \leq \delta\right).$$

On the other hand,

$$P\left(\left|\int_0^T u_s dB_s\right| \geq K, \int_0^T u_s^2 ds \leq \delta\right) = P\left(\left|\int_0^T u_s dB_s\right| \geq K, \tau = T\right)$$
$$\leq P\left(\left|\int_0^\tau u_s dB_s\right| \geq K\right)$$
$$\leq \frac{1}{K^2} E\left(\left|\int_0^\tau u_s dB_s\right|^2\right) = \frac{1}{K^2} E\left(\int_0^\tau u_s^2 ds\right) \leq \frac{\delta}{K^2},$$

which concludes the proof. $\qquad\qquad\square$

As a consequence of the above proposition, if $u^{(n)}$ is a sequence of processes in $L^2_{loc}(\mathcal{P})$ that converges to $u \in L^2_{loc}(\mathcal{P})$ in probability, that is, for all $\epsilon > 0$ and $T > 0$,

$$\lim_{n\to\infty} P\left(\left|\int_0^T (u_s^n - u_s)^2 ds\right| > \epsilon\right) = 0,$$

then the sequence $\int_0^T u_s^n dB_s$ converges in probability to $\int_0^T u_s dB_s$, for all $T > 0$.

Moreover, we can show (see Exercise 2.3) that, for all $u \in L^2_{loc}(\mathcal{P})$ and

$t \geq 0$, if $\pi = \{0 = t_0 < t_1 < \cdots < t_n = t\}$ denotes a subdivision of the interval $[0, t]$, we have that, as $|\pi| \to 0$,

$$\sum_{j=0}^{n-1} \left(\int_{t_j}^{t_{j+1}} u_s dB_s \right)^2 \xrightarrow{P} \int_0^t u_s^2 ds. \tag{2.10}$$

Therefore, $\int_0^t u_s^2 ds$ is the quadratic variation of the local martingale $\int_0^t u_s dB_s$ (see Theorem A.7.2).

2.4 Itô's Formula

Itô's stochastic integral does not follow the chain rule of classical calculus. For instance, in Example 2.1.6, we showed that

$$\int_0^t B_s dB_s = \tfrac{1}{2} B_t^2 - \tfrac{1}{2} t, \tag{2.11}$$

whereas, if x_t is a differentiable function such that $x_0 = 0$,

$$\int_0^t x_s dx_s = \int_0^t x_s x_s' ds = \tfrac{1}{2} x_t^2.$$

Equation (2.11) can also be written as

$$B_t^2 = \int_0^t 2B_s dB_s + t.$$

That is, the stochastic process B_t^2 can be expressed as the sum of the indefinite stochastic integral $\int_0^t 2B_s dB_s$ plus a differentiable function. More generally, Itô's formula below shows that any process of the form $f(B_t)$, where f is twice continuously differentiable, can be expressed as the sum of an indefinite Itô integral and a process with differentiable trajectories. This leads to the definition of an Itô process.

Denote by $L_{loc}^1(\mathcal{P})$ the space of progressively measurable processes $v = (v_t)_{t \geq 0}$ such that, for all $t > 0$,

$$P\left(\int_0^t |v_s| ds < \infty \right) = 1.$$

Definition 2.4.1 A continuous and adapted stochastic process $(X_t)_{t \geq 0}$ is called an *Itô process* if

$$X_t = X_0 + \int_0^t u_s dB_s + \int_0^t v_s ds, \tag{2.12}$$

where $u \in L_{loc}^2(\mathcal{P})$, $v \in L_{loc}^1(\mathcal{P})$, and X_0 is an $\mathcal{F}_0$-measurable variable.

Notice that $\mathcal{F}_0$-measurable random variables are constant a.s.

Using (2.10), one can easily show (see Exercise 2.4) the following property.

Proposition 2.4.2 *Let* $(X_t)_{t \geq 0}$ *and* $(Y_t)_{t \geq 0}$ *be two Itô processes of the form*

$$X_t = X_0 + \int_0^t u_s^X dB_s + \int_0^t v_s^X ds$$

and

$$Y_t = Y_0 + \int_0^t u_s^Y dB_s + \int_0^t v_s^Y ds,$$

where $u^X, u^Y \in L^2_{loc}(\mathcal{P})$ *and* $v^X, v^Y \in L^1_{loc}(\mathcal{P})$. *Let* $\pi = \{0 = t_0 < t_1 < \cdots < t_n = t\}$ *be a partition of the interval* $[0,t]$. *Then, as* $|\pi| \to 0$,

$$\sum_{j=0}^{n-1} (Y_{t_{j+1}} - Y_{t_j})(X_{t_{j+1}} - X_{t_j}) \xrightarrow{P} \int_0^t u_s^X u_s^Y ds.$$

Therefore, $\langle X, Y \rangle_t = \int_0^t u_s^X u_s^Y ds$ *is the quadratic covariation between the local martingales* $\int_0^t u_s^X dB_s$ *and* $\int_0^t u_s^Y dB_s$ *(see Definition A.7.3).*

We say that a function $f: \mathbb{R}_+ \times \mathbb{R} \to \mathbb{R}$ is of class $C^{1,2}$ if it is twice differentiable with respect to the variable $x \in \mathbb{R}$ and once differentiable with respect to $t \in \mathbb{R}_+$, with continuous partial derivatives $\partial f / \partial x$, $\partial^2 f / \partial x^2$, and $\partial f / \partial t$.

Theorem 2.4.3 (Itô's formula) *Let* $f: \mathbb{R}_+ \times \mathbb{R} \to \mathbb{R}$ *be a function of class* $C^{1,2}$. *Suppose that* X *is an Itô process of the form* (2.12). *Then, the process* $Y_t = f(t, X_t)$ *is also an Itô process, with representation*

$$Y_t = f(0, X_0) + \int_0^t \frac{\partial f}{\partial t}(s, X_s)ds + \int_0^t \frac{\partial f}{\partial x}(s, X_s) u_s dB_s$$

$$+ \int_0^t \frac{\partial f}{\partial x}(s, X_s) v_s ds + \frac{1}{2} \int_0^t \frac{\partial^2 f}{\partial x^2}(s, X_s) u_s^2 ds, \qquad (2.13)$$

which holds a.s., for all $t \geq 0$.

Remark 2.4.4 (1) Notice that all the integrals in (2.13) are well defined, that is, the processes

$$\frac{\partial f}{\partial t}(s, X_s), \quad \frac{\partial f}{\partial x}(s, X_s) v_s, \quad \text{and} \quad \frac{\partial^2 f}{\partial x^2}(s, X_s) u_s^2$$

belong to $L^1_{loc}(\mathcal{P})$, and $(\partial f/\partial x)(s, X_s) u_s$ belongs to $L^2_{loc}(\mathcal{P})$.

(2) We set $\langle X \rangle_t = \int_0^t u_s^2 ds$, that is, $\langle X \rangle_t$ is the quadratic variation of the

local martingale $\int_0^t u_s dB_s$ in the sense of (2.10). Then (2.13) can be written as

$$Y_t = f(0, X_0) + \int_0^t \frac{\partial f}{\partial t}(s, X_s) ds + \int_0^t \frac{\partial f}{\partial x}(s, X_s) dX_s + \frac{1}{2} \int_0^t \frac{\partial^2 f}{\partial x^2}(s, X_s) d\langle X \rangle_s.$$

(3) In differential notation, we write

$$dX_t = u_t dB_t + v_t dt,$$

and Itô's formula can be written as

$$df(t, X_t) = \frac{\partial f}{\partial t}(t, X_t) dt + \frac{\partial f}{\partial x}(t, X_t) dX_t + \frac{1}{2} \frac{\partial^2 f}{\partial x^2}(t, X_t) (dX_t)^2,$$

where $(dX_t)^2$ is computed using the product rule

$\times$	dB_t	dt
dB_t	dt	0
dt	0	0

Notice also that $(dX_t)^2 = u_t^2 dt = d\langle X \rangle_t$.

(4) In the particular case $u_t = 1$, $v_t = 0$, and $X_0 = 0$, the process X is the Brownian motion B, and Itô's formula has the following simple form:

$$f(t, B_t) = f(0, 0) + \int_0^t \frac{\partial f}{\partial x}(s, B_s) dB_s + \int_0^t \frac{\partial f}{\partial t}(s, B_s) ds + \frac{1}{2} \int_0^t \frac{\partial^2 f}{\partial x^2}(s, B_s) ds.$$

Proof of Theorem 2.4.3   We will give the proof only in the case where $v_t = 0$, that is,

$$X_t = X_0 + \int_0^t u_s dB_s,$$

and f does not depend on t. The extension of the proof to the general case is easy and we omit the details. We claim that

$$f(X_t) = f(X_0) + \int_0^t f'(X_s) u_s dB_s + \frac{1}{2} \int_0^t f''(X_s) u_s^2 ds.$$

By a localization argument, we may assume that $f \in C_b^2(\mathbb{R})$, $\int_0^\infty u_s^2 ds \leq N$, and $\sup_{t \geq 0} |X_t| \leq N$, for some integer $N \geq 1$ such that $|X_0| \leq N$. In fact, consider the sequence of stopping times

$$T_N = \inf \left\{ t \geq 0 : \int_0^t u_s^2 ds \geq N, \quad \text{or} \quad |X_t| \geq N \right\}.$$

Let $f_N \colon \mathbb{R} \to \mathbb{R}$ be a function in $C_0^2(\mathbb{R})$ such that $f(x) = f_N(x)$ for $|x| \le N$.
Then, if $u_t^{(N)} = u_t \mathbf{1}_{[0,T_N]}(t)$ and

$$X_t^{(N)} = X_0 + \int_0^t u_s^{(N)} dB_s,$$

by Itô's formula we get

$$f_N(X_t^{(N)}) = f_N(X_0) + \int_0^t f_N'(X_s^{(N)}) u_s^{(N)} dB_s + \frac{1}{2} \int_0^t f_N''(X_s^{(N)})(u_s^{(N)})^2 ds.$$

By Proposition 2.3.1,

$$X_t^{(N)} = X_0 + \int_0^{T_N \wedge t} u_s dB_s = X_{T_N \wedge t},$$

and

$$f(X_{T_N \wedge t}) = f(X_0) + \int_0^{T_N \wedge t} f'(X_s) u_s dB_s + \frac{1}{2} \int_0^{T_N \wedge t} f''(X_s) u_s^2 ds.$$

Then, we let $N \to \infty$ to get the result.

Consider the uniform partition $0 = t_0 < t_1 < \cdots < t_n = t$, where $t_i = it/n$.
We can write, using Taylor's formula,

$$f(X_t) - f(X_0) = \sum_{i=0}^{n-1} (f(X_{t_{i+1}}) - f(X_{t_i}))$$

$$= \sum_{i=0}^{n-1} f'(X_{t_i})(X_{t_{i+1}} - X_{t_i}) + \frac{1}{2} \sum_{i=0}^{n-1} f''(\widetilde{X}_i)(X_{t_{i+1}} - X_{t_i})^2,$$

where $\widetilde{X}_i$ is a random point between X_{t_i} and $X_{t_{i+1}}$.

It is an easy exercise to show that

$$\sum_{i=0}^{n-1} f'(X_{t_i})(X_{t_{i+1}} - X_{t_i}) \xrightarrow{L^2(\Omega)} \int_0^t f'(X_s) u_s dB_s, \tag{2.14}$$

as n tends to infinity (see Exercise 2.5).

For the second term, we write

$$\int_0^t f''(X_s)u_s^2 ds - \sum_{i=0}^{n-1} f''(\widetilde{X}_i)(X_{t_{i+1}} - X_{t_i})^2$$

$$= \sum_{i=0}^{n-1} \int_{t_i}^{t_{i+1}} (f''(X_s) - f''(X_{t_i}))u_s^2 ds$$

$$+ \sum_{i=0}^{n-1} f''(X_{t_i}) \left(\int_{t_i}^{t_{i+1}} u_s^2 ds - \left(\int_{t_i}^{t_{i+1}} u_s dB_s \right)^2 \right)$$

$$+ \sum_{i=0}^{n-1} (f''(X_{t_i}) - f''(\widetilde{X}_i)) \left(\int_{t_i}^{t_{i+1}} u_s dB_s \right)^2$$

$$=: A_1^n + A_2^n + A_3^n.$$

Then, it suffices to show that each term A_i^n converges to zero in probability. We have

$$|A_1^n| \le \sup_{|s-r| \le t/n} |f''(X_s) - f''(X_r)| \int_0^t u_s^2 ds$$

and

$$|A_3^n| \le \sup_{0 \le i \le n-1} |f''(X_{t_i}) - f''(\widetilde{X}_i)| \sum_{i=0}^{n-1} \left(\int_{t_i}^{t_{i+1}} u_s dB_s \right)^2.$$

Taking into account that f'' and X are continuous, together with Proposition 2.2.2, we obtain that both expressions converge to zero in probability as n tends to infinity.

As the sequence $\xi_i = \int_{t_i}^{t_{i+1}} u_s^2 ds - \left(\int_{t_i}^{t_{i+1}} u_s dB_s \right)^2$ is bounded and satisfies $E(\xi_i | \mathcal{F}_{t_i}) = 0$, we obtain

$$E((A_n^2)^2) = \sum_{i=0}^{n-1} E(f''(X_{t_i})^2 \xi_i^2) \le \|f''\|_\infty^2 \sum_{i=0}^{n-1} E(\xi_i^2)$$

$$\le 2\|f''\|_\infty^2 \sum_{i=0}^{n-1} E\left(\left(\int_{t_i}^{t_{i+1}} u_s^2 ds \right)^2 + \left(\int_{t_i}^{t_{i+1}} u_s dB_s \right)^4 \right)$$

$$\le 2\|f''\|_\infty^2 E\left(N \sup_i \int_{t_i}^{t_{i+1}} u_s^2 ds \right.$$

$$\left. + \sup_i |X_{t_{i+1}} - X_{t_i}|^2 \sum_{i=0}^{n-1} \left(\int_{t_i}^{t_{i+1}} u_s dB_s \right)^2 \right),$$

which converges to zero as $n \to \infty$. This completes the proof. $\square$

Example 2.4.5 (1) If $f(x) = x^2$ and $X_t = B_t$, we obtain

$$B_t^2 = 2 \int_0^t B_s dB_s + t,$$

because $f'(x) = 2x$ and $f''(x) = 2$.

(2) If $f(x) = x^3$ and $X_t = B_t$, we obtain

$$B_t^3 = 3 \int_0^t B_s^2 dB_s + 3 \int_0^t B_s ds,$$

because $f'(x) = 3x^2$ and $f''(x) = 6x$. More generally, if $n \geq 2$ is a natural number,

$$B_t^n = n \int_0^t B_s^{n-1} dB_s + \frac{n(n-1)}{2} \int_0^t B_s^{n-2} ds.$$

(3) If $f(t, x) = \exp(ax - a^2 t/2)$, $X_t = B_t$, and $Y_t = \exp(aB_t - a^2 t/2)$, we obtain

$$Y_t = 1 + a \int_0^t Y_s dB_s$$

because

$$\frac{\partial f}{\partial t} + \frac{1}{2}\frac{\partial^2 f}{\partial x^2} = 0. \tag{2.15}$$

In particular, the solution to the stochastic differential equation

$$dY_t = aY_t dB_t, \quad Y_0 = 1,$$

is not $Y_t = \exp(aB_t)$ but $Y_t = \exp(aB_t - a^2 t/2)$.

(4) If a function f of class $C^{1,2}$ satisfies equality (2.15) then,

$$f(t, B_t) = f(0, 0) + \int_0^t \frac{\partial f}{\partial x}(s, B_s) dB_s.$$

This implies that $f(t, B_t)$ is a continuous local martingale. The process is a square integrable martingale if for all $t \geq 0$

$$E\left(\int_0^t \left(\frac{\partial f}{\partial x}(s, B_s) \right)^2 ds \right) < \infty.$$

2.5 Tanaka's Formula

Consider the two-parameter random field $(L_t^x)_{t \geq 0, x \in \mathbb{R}}$ defined as

$$L_t^x = \lim_{\epsilon \downarrow 0} \frac{1}{2\epsilon} \ell\{0 \leq s \leq t : |B_s - x| \leq \epsilon\}, \tag{2.16}$$

where ℓ denotes the Lebesgue measure. Lévy (1948) proved that this limit exists and is finite. Moreover, there exists a version of this process which is continuous in both variables (t, x). This random field is called the Brownian local time, whose name is justified by the following fact. For any Borel set $A \in \mathcal{B}(\mathbb{R})$, we define the occupation measure of the Brownian motion up to time $t \geq 0$ as

$$\mu_t(A) = \int_0^t \mathbf{1}_{\{B_s \in A\}} ds.$$

Then, this measure is absolutely continuous with respect to the Lebesgue measure, and its density coincides with the Brownian local time L_t^x defined by (2.16) (see for instance Yor, 1986). That is, for any bounded or nonnegative measurable function $f : \mathbb{R} \to \mathbb{R}$,

$$\int_0^t f(B_s) ds = \int_{\mathbb{R}} f(x) L_t^x dx. \tag{2.17}$$

Formally, we can write $L_t^x = \int_0^t \delta_x(B_s) ds$, that is, the local time is the time spent by the Brownian motion at x in the time interval $[0, t]$.

Tanaka's formula (Tanaka, 1963) gives a representation of the Brownian local time.

Theorem 2.5.1 (Tanaka's formula)   *Almost surely, for any $x \in \mathbb{R}$ and $t > 0$,*

$$\tfrac{1}{2} L_t^x = (B_t - x)_+ - (-x)_+ - \int_0^t \mathbf{1}_{\{B_s > x\}} dB_s \tag{2.18}$$

and

$$\tfrac{1}{2} L_t^x = (B_t - x)_- - (-x)_- + \int_0^t \mathbf{1}_{\{B_s < x\}} dB_s. \tag{2.19}$$

Trotter (1958) showed the existence of the local time by first showing that the right-hand side of (2.18) admits a continuous modification in (t, x) and then using Tanaka's formula. See also Karatzas and Shreve (1998, Theorem 6.11) for a proof of Trotter's theorem.

Proof of Theorem 2.5.1   We consider an approximation of the Dirac delta function $\delta_0(y)$ by a sequence of probability density functions. More specifically, we set

$$f_n(y) = n f(ny),$$

where f is the $C^\infty(\mathbb{R})$ function given by

$$f(y) = \begin{cases} c\exp(1/((y-1)^2 - 1)) & \text{if } 0 < y < 2, \\ 0 & \text{otherwise;} \end{cases}$$

here the constant c is chosen in such a way that $\int_{\mathbb{R}} f(y)dy = 1$. We also consider the sequence of functions

$$u_n(y) = \int_{-\infty}^{y}\int_{-\infty}^{w} f_n(z-x)dzdw, \quad y \in \mathbb{R}, \ n \geq 1.$$

We observe that $u_n'(y) = \int_{-\infty}^{y} f_n(z-x)dz$, which implies that for all $y \in \mathbb{R}$,

$$\lim_{n\to\infty} u_n'(y) = \mathbf{1}_{(x,\infty)}(y) \quad \text{and} \quad \lim_{n\to\infty} u_n(y) = (y-x)_+.$$

Applying Itô's formula (Theorem 2.4.3), we obtain, for all $t \geq 0$,

$$u_n(B_t) = u_n(0) + \int_0^t u_n'(B_s)dB_s + \frac{1}{2}\int_0^t f_n(B_s - x)ds. \tag{2.20}$$

Now, from (2.17) we get

$$\int_0^t f_n(B_s - x)ds = \int_{\mathbb{R}} f_n(y-x)L_t^y dy,$$

which, by the continuity of the local time, converges to L_t^x almost surely as $n \to \infty$. Moreover,

$$E\left(\left|\int_0^t u_n'(B_s)dB_s - \int_0^t \mathbf{1}_{(x,\infty)}(B_s)dB_s\right|^2\right) = E\left(\int_0^t \left|u_n'(B_s) - \mathbf{1}_{(x,\infty)}(B_s)\right|^2 ds\right)$$

$$\leq \int_0^t P\left(|B_s - x| \leq \frac{2}{n}\right)ds,$$

which converges to zero as $n \to \infty$. Therefore, equation (2.18) for each fixed $t > 0$ follows on letting $n \to \infty$ in (2.20). Because of the continuity of the processes, we obtain the result almost surely for all $t > 0$ and $x \in \mathbb{R}$. Equation (2.19) can be deduced from (2.18), taking into account that L_t^x coincides with the local time at $-x$ of the Brownian motion $-B$. $\qquad\square$

Corollary 2.5.2 *Almost surely, for any $x \in \mathbb{R}$ and $t > 0$,*

$$L_t^x = |B_t - x| - |x| - \int_0^t \text{sign}(B_s - x)\,dB_s,$$

where $\text{sign}(x) = \mathbf{1}_{\{x>0\}} - \mathbf{1}_{\{x<0\}}$.

2.6 Multidimensional Version of Itô's Formula

Itô's formula can be extended to a multidimensional setting as follows. Suppose that $B = (B_t)_{t\geq0}$, $B_t = (B_t^1, B_t^2, \ldots, B_t^m)$, is an m-dimensional Brownian motion.

Definition 2.6.1 An n-dimensional continuous and adapted process $(X_t)_{t\geq0}$ is called a *multidimensional Itô process* if

$$X_t = X_0 + \int_0^t u_s dB_s + \int_0^t v_s ds, \qquad (2.21)$$

where $v = (v_t)_{t\geq0}$ is an n-dimensional process, $u = (u_t)_{t\geq0}$ is a process with values in the set of $n \times m$ matrices, and we assume that the components of u belong to $L_{loc}^2(\mathcal{P})$ and those of v belong to $L_{loc}^1(\mathcal{P})$.

We say that a function $f: \mathbb{R}_+ \times \mathbb{R}^n \to \mathbb{R}$ is of class $C^{1,2}$ if the partial derivatives $\partial f/\partial t$, $\partial f/\partial x_i$, and $\partial f/\partial x_i \partial x_j$, $1 \leq i, j \leq n$, exist and are continuous.

Theorem 2.6.2 (Multidimensional version of Itô's formula) *Suppose that X is a multidimensional Itô process of the form* (2.21). *Let* $f: \mathbb{R}_+ \times \mathbb{R}^n \to \mathbb{R}$ *be a function of class $C^{1,2}$. Then, the process $Y_t = f(t, X_t)$ is also a multidimensional Itô process with representation*

$$Y_t = f(0, X_0) + \int_0^t \frac{\partial f}{\partial t}(s, X_s)ds + \sum_{i=1}^n \int_0^t \frac{\partial f}{\partial x_i}(s, X_s)dX_s^i$$
$$+ \frac{1}{2} \sum_{i,j=1}^n \int_0^t \frac{\partial^2 f}{\partial x_i \partial x_j}(s, X_s)(u_s u_s^T)_{ij} ds$$

almost surely, for all $t \geq 0$.

Proof The proof follows the same lines as that for the one-dimensional Itô's formula (Theorem 2.4.3) using Exercise 2.10. □

Remark 2.6.3 (1) Set

$$\langle X^i, X^j \rangle_t = \int_0^t (u_s u_s^T)_{ij} ds = \int_0^t \sum_{k=1}^m (u_s)_{ik}(u_s)_{jk} ds.$$

That is, $\langle X^i, X^j \rangle_t$ is the quadratic covariation of the local martingales $\sum_{k=1}^m \int_0^t (u_s)_{ik} dB_s^k$ and $\sum_{k=1}^m \int_0^t (u_s)_{jk} dB_s^k$. Then, Itô's formula can be written in the form

$$Y_t = f(0, X_0) + \int_0^t \frac{\partial f}{\partial t}(s, X_s)ds + \sum_{i=1}^n \int_0^t \frac{\partial f}{\partial x_i}(s, X_s)dX_s^i$$

$$+ \frac{1}{2} \sum_{i,j=1}^n \int_0^t \frac{\partial^2 f}{\partial x_i \partial x_j}(s, X_s)d\langle X^i, X^j \rangle_s.$$

(2) In differential notation, the multidimensional Itô's formula can be written as

$$dY_t = \frac{\partial f}{\partial t}(t, X_t)dt + \nabla f(t, X_t)dX_t + \frac{1}{2} \sum_{i,j=1}^n \frac{\partial^2 f}{\partial x_i \partial x_j}(t, X_t)dX_t^i dX_t^j.$$

The product of differentials $dX_t^i dX_t^j$ is computed by means of the product rules $dB_t^i dt = 0$, $(dt)^2 = 0$, and

$$dB_t^i dB_t^j = \begin{cases} 0 & \text{if } i \neq j, \\ dt & \text{if } i = j. \end{cases}$$

(3) As an application of the multidimensional Itô's formula we can deduce the following integration-by-parts formula. Suppose that $(X_t)_{t \geq 0}$ and $(Y_t)_{t \geq 0}$ are one-dimensional Itô processes. Then

$$X_t Y_t = X_0 Y_0 + \int_0^t X_s dY_s + \int_0^t Y_s dX_s + \int_0^t d\langle X, Y \rangle_s. \qquad (2.22)$$

The multidimensional Itô's formula leads to the following result, whose proof is left an exercise (Exercise 2.11).

Proposition 2.6.4 *Let $f: \mathbb{R}_+ \times \mathbb{R}^d \to \mathbb{R}$ be a function of class $C^{1,2}$, and let B be a d-dimensional Brownian motion. Then, the process*

$$X_t = f(t, B_t) - \int_0^t \left(\tfrac{1}{2}\Delta f(s, B_s) + \frac{\partial f}{\partial t}(s, B_s) \right) ds$$

is a local martingale. If, moreover,

$$\sum_{i=1}^d \left(\frac{\partial f}{\partial x_i}(t, x) \right)^2 \leq K_2(t) \, \exp(K_1 |x|^\beta),$$

for some constants $K_1 \geq 0$, $\beta \in [0, 2)$, and a function $K_2(t) \geq 0$ such that $\int_0^T K_2(t)dt < \infty$ for all $T > 0$, then $(X_t)_{t \geq 0}$ is a martingale.

2.7 Stratonovich Integral

Itô's integral was defined as the limit in $L^2(\Omega)$ or in probability of forward Riemann sums. If we consider symmetric Riemann sums defined by taking the value of the integrand in the middle point of each interval, we obtain a different type of integral, called the Stratonovich integral. In the next proposition we define the Stratonovich integral of one Itô process with respect to another.

Proposition 2.7.1 *Let $(X_t)_{t \geq 0}$ and $(Y_t)_{t \geq 0}$ be two Itô processes of the form*

$$X_t = X_0 + \int_0^t u_s^X dB_s + \int_0^t v_s^X ds$$

and

$$Y_t = Y_0 + \int_0^t u_s^Y dB_s + \int_0^t v_s^Y ds,$$

where $u^X, u^Y \in L^2_{loc}(\mathcal{P})$ and $v^X, v^Y \in L^1_{loc}(\mathcal{P})$. Let $\pi = \{0 = t_0 < t_1 < \cdots < t_n = t\}$ be a partition of the interval $[0, t]$. Then, as $|\pi| \to 0$,

$$\sum_{j=0}^{n-1} \tfrac{1}{2}(Y_{t_j} + Y_{t_{j+1}})(X_{t_{j+1}} - X_{t_j}) \xrightarrow{P} \int_0^t Y_s dX_s + \tfrac{1}{2}\langle X, Y \rangle_t.$$

This limit is called the Stratonovich integral of Y with respect to X and is denoted by $\int_0^t Y_s \circ dX_s$.

Proof   The result follows easily from the decomposition

$$\tfrac{1}{2}(Y_{t_j} + Y_{t_{j+1}})(X_{t_{j+1}} - X_{t_j}) = Y_{t_j}(X_{t_{j+1}} - X_{t_j}) + \tfrac{1}{2}(Y_{t_{j+1}} - Y_{t_j})(X_{t_{j+1}} - X_{t_j})$$

and Proposition 2.4.2.   □

The Stratonovich integral follows the rules of classical calculus. That is, if $f \in C^3(\mathbb{R})$, then

$$f(X_t) = f(X_0) + \int_0^t f'(X_s) \circ dX_s.$$

Indeed, by Itô's formula (Theorem 2.4.3) and Proposition 2.7.1,

$$f(X_t) = f(X_0) + \int_0^t f'(X_s) dX_s + \tfrac{1}{2} \int_0^t f''(X_s) d\langle X \rangle_s$$

$$= f(X_0) + \int_0^t f'(X_s) \circ dX_s - \tfrac{1}{2}\langle f'(X), X \rangle_t + \tfrac{1}{2} \int_0^t f''(X_s) d\langle X \rangle_s.$$

Furthermore, again by Itô's formula, $\langle f'(X), X \rangle_t$ is the quadratic covariation between the martingales $\int_0^t f''(X_s) u_s dB_s$ and $\int_0^t u_s dB_s$, which coincides with $\int_0^t f''(X_s) d\langle X \rangle_s$.

2.8 Backward Stochastic Integral

Let $B = (B_t)_{t \geq 0}$ be a Brownian motion. Fix $T > 0$. For any $t \in [0, T]$ we denote by $\hat{\mathcal{F}}_t$ the σ-field generated by the random variables $\{B_T - B_s, t \leq s \leq T\}$ and the sets of probability zero. Notice that $(\hat{\mathcal{F}}_t)_{t \in [0, T]}$ is a decreasing family of σ-fields.

A stochastic process $u = (u_t)_{t \in [0, T}$ is called *backward predictable* if for all $t \in [0, T]$, $u_t(\omega)$ restricted to $\Omega \times [t, T]$ is measurable with respect to the σ-field $\hat{\mathcal{F}}_t \times \mathcal{B}([t, T])$. We denote by $L_T^2(\hat{\mathcal{P}})$ the space of backward predictable processes u such that $E\left(\int_0^T u_s^2 ds\right) < \infty$.

For a process $u \in L_T^2(\hat{\mathcal{P}})$ we define the backward Itô stochastic integral of u as the following limit in $L^2(\Omega)$:

$$\int_0^T u_t \hat{d}B_t = \lim_{n \to \infty} \sum_{j=0}^{n-2} \left(\frac{n}{T} \int_{(j+1)T/n}^{(j+2)T/n} u_s ds\right)(B_{(j+1)T/n} - B_{jT/n}).$$

If the process u is adapted to the backward filtration $\hat{\mathcal{F}}_t$ and is continuous in $L^2(\Omega)$ then the backward Itô stochastic integral is the limit of backward Riemann sums, that is,

$$\int_0^T u_t \hat{d}B_t = \lim_{n \to \infty} \sum_{j=0}^{n-1} u_{(j+1)T/n}(B_{(j+1)T/n} - B_{jT/n}).$$

The process $\hat{B}_t = B_T - B_{T-t}$, $t \in [0, T]$, is a Brownian motion with natural filtration $\hat{\mathcal{F}}_{T-t}$. Then, one can easily show (see Exercise 2.12) that for any process $u \in L_T^2(\hat{\mathcal{P}})$,

$$\int_0^T u_t \hat{d}B_t = \int_0^T u_{T-t} d\hat{B}_t.$$

As a consequence, the backward Itô integral has the properties similar to those of the Itô integral.

2.9 Integral Representation Theorem

Consider a process u in the space $L^2_\infty(\mathcal{P})$. We know that the indefinite stochastic integral

$$X_t = \int_0^t u_s dB_s$$

is a martingale with respect to the filtration $(\mathcal{F}_t)_{t \geq 0}$. The aim of this subsection is to show that any square integrable martingale is of this form. We start with the integral representation of square integrable random variables.

Theorem 2.9.1 (Integral representation theorem)   *Fix $T > 0$ and let $F \in L^2(\Omega, \mathcal{F}_T, \mathrm{P})$. Then there exists a unique process u in the space $L^2_T(\mathcal{P})$ such that*

$$F = \mathrm{E}(F) + \int_0^T u_s dB_s.$$

Proof   Suppose first that

$$F = \exp\left(\int_0^T h_s dB_s - \frac{1}{2} \int_0^T h_s^2 ds \right), \tag{2.23}$$

where $h \in L^2([0, T])$. Define

$$Y_t = \exp\left(\int_0^t h_s dB_s - \frac{1}{2} \int_0^t h_s^2 ds \right), \quad 0 \leq t \leq T.$$

By Itô's formula applied to the function $f(x) = e^x$ and the process $X_t = \int_0^t h_s dB_s - \frac{1}{2} \int_0^t h_s^2 ds$, we obtain, for all $t \in [0, T]$,

$$Y_t = 1 + \int_0^t Y_s h_s dB_s.$$

Hence,

$$F = 1 + \int_0^T Y_s h_s dB_s$$

and we get the desired representation because $\mathrm{E}(F) = 1$ and $Yh \in L^2_\infty(\mathcal{P})$. By linearity, the representation holds for linear combinations of exponentials of the form (2.23).

In the general case, any random variable $F \in L^2(\Omega, \mathcal{F}_T, \mathrm{P})$ can be approximated in $L^2(\Omega)$ by a sequence F_n of linear combinations of exponentials of the form (2.23). Then, we have

$$F_n = \mathrm{E}(F_n) + \int_0^T u_s^{(n)} dB_s.$$

By the isometry of the stochastic integral,

$$E((F_n - F_m)^2) \geq \text{Var}(F_n - F_m) = E\left(\left(\int_0^T (u_s^{(n)} - u_s^{(m)})dB_s\right)^2\right)$$

$$= E\left(\int_0^T (u_s^{(n)} - u_s^{(m)})^2 ds\right).$$

Hence, $u^{(n)}$ is a Cauchy sequence in $L_T^2(\mathcal{P})$ and it converges to a process u in $L_T^2(\mathcal{P})$. Applying again the isometry property, and taking into account that $E(F_n)$ converges to $E(F)$, we obtain

$$F = \lim_{n\to\infty} F_n = \lim_{n\to\infty} \left(E(F_n) + \int_0^T u_s^{(n)} dB_s\right)$$

$$= E(F) + \int_0^T u_s dB_s.$$

Finally, uniqueness also follows from the isometry property. Indeed, suppose that $u^{(1)}$ and $u^{(2)}$ are processes in $L_T^2(\mathcal{P})$ such that

$$F = E(F) + \int_0^T u_s^{(1)} dB_s = E(F) + \int_0^T u_s^{(2)} dB_s.$$

Then

$$0 = E\left(\left(\int_0^T (u_s^{(1)} - u_s^{(2)})dB_s\right)^2\right) = E\left(\int_0^T (u_s^{(1)} - u_s^{(2)})^2 ds\right),$$

and hence $u_s^{(1)}(\omega) = u_s^{(2)}(\omega)$ for almost all $(\omega, s) \in \Omega \times [0, T]$. □

Corollary 2.9.2 (Martingale representation theorem) *Suppose that $M = (M_t)_{t\geq 0}$ is a square integrable martingale with respect to $(\mathcal{F}_t)_{t\geq 0}$. Then there exists a unique process $u \in L_\infty^2(\mathcal{P})$ such that*

$$M_t = E(M_0) + \int_0^t u_s dB_s.$$

In particular, M has a continuous version.

Example 2.9.3 We want to find the integral representation of $F = B_T^3$. By Itô's formula,

$$B_T^3 = \int_0^T 3B_t^2 dB_t + 3\int_0^T B_t dt$$

and, using the integration-by-parts formula (2.22) yields

$$\int_0^T B_t dt = TB_T - \int_0^T t dB_t = \int_0^T (T - t)dB_t.$$

So, we obtain the representation

$$B_T^3 = \int_0^T 3(B_t^2 + (T - t))dB_t.$$

2.10 Girsanov's Theorem

Girsanov's theorem says that a Brownian motion with drift $(B_t + \lambda t)_{t\in[0,T]}$ can be seen as a Brownian motion without drift under a suitable probability measure.

Suppose that L is a nonnegative random variable on a probability space $(\Omega, \mathcal{F}, P)$ such that $E(L) = 1$. Then $Q(A) = E(\mathbf{1}_A L)$ defines a new probability on $(\Omega, \mathcal{F})$. In fact, Q is a σ-additive measure such that $Q(\Omega) = E(L) = 1$. We say that L is the density of Q with respect to P, and we write $dQ/dP = L$. The probability Q is absolutely continuous with respect to P, that is, for any $A \in \mathcal{F}$,

$$P(A) = 0 \quad \Longrightarrow \quad Q(A) = 0.$$

If L is strictly positive P-almost surely, then the probabilities P and Q are equivalent (that is, mutually absolutely continuous). This means that for any $A \in \mathcal{F}$,

$$P(A) = 0 \quad \Longleftrightarrow \quad Q(A) = 0.$$

The next example is a simple version of Girsanov's theorem.

Example 2.10.1   Let X be a random variable with distribution $N(m, \sigma^2)$. Consider the random variable

$$L = \exp\left(-\frac{m}{\sigma^2}X + \frac{m^2}{2\sigma^2}\right),$$

which satisfies $E(L) = 1$. Suppose that Q has density L with respect to P. On the probability space $(\Omega, \mathcal{F}, Q)$, the variable X has characteristic function

$$E_Q(e^{itX}) = E(e^{itX}L) = \frac{1}{\sqrt{2\pi\sigma^2}} \int_{-\infty}^{\infty} \exp\left(-\frac{(x-m)^2}{2\sigma^2} - \frac{mx}{\sigma^2} + \frac{m^2}{2\sigma^2} + itx\right)dx$$

$$= \frac{1}{\sqrt{2\pi\sigma^2}} \int_{-\infty}^{\infty} \exp\left(-\frac{x^2}{2\sigma^2} + itx\right)dx = \exp\left(-\frac{\sigma^2 t^2}{2}\right),$$

so, X has distribution $N(0, \sigma^2)$.

We now go back to the case where $(B_t)_{t\geq0}$ is a Brownian motion on a probability space $(\Omega, \mathcal{F}, P)$. Given a process $\theta \in L_T^2(\mathcal{P})$, consider the local martingale

$$L_t = \exp\left(\int_0^t \theta_s dB_s - \frac{1}{2} \int_0^t \theta_s^2 ds \right), \quad 0 \le t \le T,$$

which satisfies the linear stochastic differential equation

$$L_t = 1 + \int_0^t \theta_s L_s dB_s, \quad 0 \le t \le T.$$

The next lemma is due to Novikov (1972), and a proof can be found in Baudoin (2014, Lemma 5.76).

Lemma 2.10.2 (Novikov's condition)   *If*

$$E\left(\exp\left(\frac{1}{2} \int_0^T \theta_s^2 ds \right) \right) < \infty \tag{2.24}$$

then $(L_t)_{0 \le t \le T}$ *is a martingale.*

Thus, as a consequence of Lemma 2.10.2, we have:

(1) The random variable L_T is a density in the probability space $(\Omega, \mathcal{F}_T, P)$ and defines a probability Q such that

$$L_T = \frac{dQ}{dP}.$$

(2) For any $t \ge 0$,

$$L_t = \frac{dQ}{dP}\bigg|_{\mathcal{F}_t}. \tag{2.25}$$

In fact, if $A \in \mathcal{F}_t$, we have

$$\begin{aligned} Q(A) &= E(\mathbf{1}_A L_T) = E(E(\mathbf{1}_A L_T | \mathcal{F}_t)) \\ &= E(\mathbf{1}_A E(L_T | \mathcal{F}_t)) \\ &= E(\mathbf{1}_A L_t). \end{aligned}$$

Girsanov's theorem was first proved by Cameron and Martin (1944) for non-random integrands, and then extended by Girsanov (1960).

Theorem 2.10.3 (Girsanov's theorem)   *Suppose that θ satisfies the Novikov condition (2.24). Then, on the probability space $(\Omega, \mathcal{F}_T, Q)$, the stochastic process*

$$W_t = B_t - \int_0^t \theta_s ds,$$

is a Brownian motion on $[0, T]$.

Proof It is enough to show that on the probability space $(\Omega, \mathcal{F}_T, Q)$, for all $s < t \leq T$, the increment $W_t - W_s$ is independent of $\mathcal{F}_s$ and has the normal distribution $N(0, t - s)$.

These properties follow from the relation, for all $s < t \leq T$, $A \in \mathcal{F}_s$, $\lambda \in \mathbb{R}$,

$$E_Q(\mathbf{1}_A \exp(i\lambda(W_t - W_s))) = Q(A) \exp\left(-\frac{\lambda^2}{2}(t - s)\right). \tag{2.26}$$

In order to show (2.26) we write, using (2.25),

$$E_Q(\mathbf{1}_A \exp(i\lambda(W_t - W_s))) = E(\mathbf{1}_A \exp(i\lambda(W_t - W_s)) L_t)$$
$$= E(\mathbf{1}_A L_s \Psi_{s,t}) \exp\left(-\frac{\lambda^2}{2}(t - s)\right),$$

where

$$\Psi_{s,t} = \exp\left(\int_s^t (i\lambda + \theta_v) dB_v - \frac{1}{2}\int_s^t (i\lambda + \theta_v)^2 dv\right).$$

Then, the desired result follows from

$$E(\Psi_{s,t}|\mathcal{F}_s) = 1.$$

$\square$

Fix $\theta \in \mathbb{R}$. Recall that under Q, given by

$$\left.\frac{dQ}{dP}\right|_{\mathcal{F}_t} = L_t := \exp\left(\theta B_t - \frac{\theta^2}{2}t\right), \tag{2.27}$$

$(B_t)_{t \geq 0}$ is a Brownian motion with drift θt. As an application, in the next proposition we compute the distribution of the hitting time for a Brownian motion with drift.

Proposition 2.10.4 *Set $\tau_a = \inf\{t \geq 0, B_t = a\}$, where $a \neq 0$. Let Q be defined by (2.27). Then, with respect to Q, the random variable τ_a has probability density*

$$f(s) = \frac{|a|}{\sqrt{2\pi s^3}} \exp\left(-\frac{(a - \theta s)^2}{2s}\right), \quad s > 0. \tag{2.28}$$

Proof For any $t \geq 0$, the event $\{\tau_a \leq t\}$ belongs to the σ-field $\mathcal{F}_{\tau_a \wedge t}$ because, for any $s \geq 0$,

$$\{\tau_a \leq t\} \cap \{\tau_a \wedge t \leq s\} = \{\tau_a \leq t\} \cap \{\tau_a \leq s\}$$
$$= \{\tau_a \leq t \wedge s\} \in \mathcal{F}_{s \wedge t} \subset \mathcal{F}_s.$$

Consequently, using the optional stopping theorem (Theorem A.7.4) yields

$$Q(\tau_a \leq t) = E(1_{\{\tau_a \leq t\}} L_t) = E(1_{\{\tau_a \leq t\}} E(L_t | \mathcal{F}_{\tau_a \wedge t}))$$
$$= E(1_{\{\tau_a \leq t\}} L_{\tau_a \wedge t}) = E(1_{\{\tau_a \leq t\}} L_{\tau_a})$$
$$= E\left(1_{\{\tau_a \leq t\}} \exp\left(\theta a - \tfrac{1}{2}\theta^2 \tau_a\right)\right)$$
$$= \int_0^t \exp\left(\theta a - \tfrac{1}{2}\theta^2 s\right) f(s)\, ds,$$

where f is the density of the random variable τ_a. We know by (1.11) that

$$f(s) = \frac{|a|}{\sqrt{2\pi s^3}} \exp\left(-\frac{a^2}{2s}\right).$$

Hence, with respect to Q the random variable τ_a has the probability density given by (2.28). □

Letting $t \uparrow \infty$ in the above proof, we obtain

$$Q(\tau_a < \infty) = \exp(\theta a)\, E\left(\exp\left(-\tfrac{1}{2}\theta^2 \tau_a\right)\right) = \exp(\theta a - |\theta a|).$$

If $\theta = 0$ (Brownian motion without drift), the probability of reaching the level a is one. If $\theta a > 0$ (the drift θ and the level a have the same sign) this probability is also one. If $\theta a < 0$ (the drift θ and the level a have opposite signs) this probability is $\exp(2\theta a)$.

Exercises

2.1 Show that the limit (2.2) does not depend on the approximating sequence $u^{(n)}$.

2.2 Let $u \in \mathcal{E}$ and $t \geq 0$. Consider a partition $\pi = \{0 = t_0 < t_1 < \cdots < t_n = t\}$. Show that, as $|\pi| \to 0$,

$$\sum_{j=0}^{n-1} \left(\int_{t_j}^{t_{j+1}} u_s\, dB_s\right)^2 \xrightarrow{L^1(\Omega)} \int_0^t u_s^2\, ds.$$

2.3 Let $u \in L^2_{loc}(\mathcal{P})$ and $t \geq 0$. Consider a partition $\pi = \{0 = t_0 < t_1 < \cdots < t_n = t\}$. Show that, as $|\pi| \to 0$,

$$\sum_{j=0}^{n-1} \left(\int_{t_j}^{t_{j+1}} u_s\, dB_s\right)^2 \xrightarrow{P} \int_0^t u_s^2\, ds.$$

2.4 Show Proposition 2.4.2.

2.5   Let $f \in C_b^2(\mathbb{R})$ and

$$X_t = X_0 + \int_0^t u_s dB_s,$$

where $u \in L_{loc}^2(\mathcal{P})$. Show that, as $n \to \infty$,

$$\sum_{i=0}^{n-1} f'(X_{t_i})(X_{t_{i+1}} - X_{t_i}) \overset{L^2(\Omega)}{\longrightarrow} \int_0^t f'(X_s)u_s dB_s,$$

where $t_i = it/n$, $0 \le i \le n$.

2.6   Use Itô's formula in order to compute $E(B_t^4)$. Do it again for $E(B_t^6)$. Write a recurrence formula to compute $E(B_t^n)$ in terms of $E(B_t^{n-2})$.

2.7   Show that, for any $x \in \mathbb{R}$, the process

$$X_t = \int_0^t \operatorname{sign}(B_s - x)dB_s, \quad t \ge 0,$$

is a Brownian motion.

Hint: Apply Itô's formula to the process $\exp(i\lambda(X_t - X_s))$.

2.8   Let B be a Brownian motion. Using Itô's formula, check whether the following processes are martingales:

(a) $X_t^{(1)} = B_t^3 - 3tB_t$.

(b) $X_t^{(2)} = t^2 B_t - 2 \int_0^t sB_s ds$.

(c) $X_t^{(3)} = e^{t/2} \cos B_t$.

(d) $X_t^{(4)} = e^{t/2} \sin B_t$.

(e) $X_t^{(5)} = (B_t + t)\exp(-B_t - \frac{1}{2}t)$.

2.9   Let B_1 and B_2 be two independent Brownian motions. Show that $B_1(t)B_2(t)$ is a martingale.

2.10 Let B be an m-dimensional Brownian motion. Consider a partition $\pi = \{0 = t_0 < t_1 < \cdots < t_n = t\}$ of the interval $[0, t]$. Show that, if $i \ne j$, as $|\pi| \to 0$ we have

$$\sum_{k=0}^{n-1}(B_{t_{k+1}}^i - B_{t_k}^i)(B_{t_{k+1}}^j - B_{t_k}^j) \overset{L^2(\Omega)}{\longrightarrow} 0.$$

2.11 Prove Proposition 2.6.4.

2.12 Using the notation of Section 2.8, show that, for any process $u \in L_T^2(\hat{\mathcal{P}})$,

$$\int_0^T u_t dB_t = \int_0^T u_{T-t} d\hat{B}_t.$$

2.13 Find the stochastic integral representation on the time interval $[0, T]$ of the following random variables:

$$B_T, \quad B_T^2, \quad e^{B_T}, \quad \int_0^T B_t dt, \quad B_T^3, \quad \sin B_T, \quad \int_0^T tB_t^2 dt.$$

2.14 Let $p(t, x) = 1/\sqrt{1-t}\exp(-x^2/(2(1-t)))$, for $0 \leq t < 1$, $x \in \mathbb{R}$, and $p(1, x) = 0$. Define $M_t = p(t, B_t)$, where $(B_t)_{t\in[0,1]}$ is a Brownian motion.

(a) Show that, for each $0 \leq t < 1$,

$$M_t = M_0 + \int_0^t \frac{\partial p}{\partial x}(s, B_s) dB_s.$$

(b) Set $H_t = (\partial p/\partial x)(t, B_t)$. Show that $\int_0^1 H_t^2 dt < \infty$ almost surely, but

$$\mathrm{E}\left(\int_0^1 H_t^2 dt \right) = \infty.$$

2.15 Let B and $\tilde{B}$ be two independent Brownian motions. Fix $t > 0$ and consider the Brownian motion $\hat{B} = e^{-t}B + \sqrt{1 - e^{-2t}}\tilde{B}$. Consider two random variables X, Y taking values in the interval $[a, b]$, where $0 < a < b$, such that X is $\hat{B}$-measurable and Y is B-measurable. Fix $p > 1$ and set $q = e^{2t}(p - 1) + 1$. Let q' be such that $1/q + 1/q' = 1$. Show that

$$\mathrm{E}(XY) \leq \|X\|_p\|Y\|_{q'}.$$

Hint: Consider the stochastic integral representations

$$X^p = \mathrm{E}(X^p) + \int_0^\infty \varphi_s d\hat{B}_s,$$

$$Y^{q'} = \mathrm{E}(Y^{q'}) + \int_0^\infty \psi_s dB_s,$$

and apply Itô's formula to the function $f(x, y) = x^{1/p}y^{1/q'}$ and to the martingales

$$M_t = \mathrm{E}(X^p) + \int_0^t \varphi_s d\hat{B}_s,$$

$$N_t = \mathrm{E}(Y^{q'}) + \int_0^t \psi_s dB_s.$$

3

Derivative and Divergence Operators

The Malliavin calculus is a differential calculus on a Gaussian probability space. In this chapter we introduce the derivative and divergence operators when the underlying process is a Brownian motion $(B_t)_{t\geq 0}$.

3.1 Finite-Dimensional Case

We consider first the finite-dimensional case. That is, the probability space $(\Omega, \mathcal{F}, P)$ is such that $\Omega = \mathbb{R}^n$, $\mathcal{F} = \mathcal{B}(\mathbb{R}^n)$ is the Borel σ-field of $\mathbb{R}^n$, and P is the standard Gaussian probability with density $p(x) = (2\pi)^{-n/2} e^{-|x|^2/2}$. In this framework we consider two differential operators. The first is the *derivative operator*, which is simply the gradient of a differentiable function $F \colon \mathbb{R}^n \to \mathbb{R}$:

$$\nabla F = \left(\frac{\partial F}{\partial x_1}, \ldots, \frac{\partial F}{\partial x_n}\right).$$

The second differential operator is the *divergence operator* and is defined on differentiable vector-valued functions $u \colon \mathbb{R}^n \to \mathbb{R}^n$ as follows:

$$\delta(u) = \sum_{i=1}^{n} \left(u_i x_i - \frac{\partial u_i}{\partial x_i}\right) = \langle u, x \rangle - \text{div } u.$$

It turns out that δ is the adjoint of the derivative operator with respect to the Gaussian measure P. This is the content of the next proposition.

Proposition 3.1.1 *The operator δ is the adjoint of ∇; that is,*

$$E(\langle u, \nabla F \rangle) = E(F \delta(u))$$

if $F \colon \mathbb{R}^n \to \mathbb{R}$ and $u \colon \mathbb{R}^n \to \mathbb{R}^n$ are continuously differentiable functions which, together with their partial derivatives, have at most polynomial growth.

Proof Integrating by parts, and using $\partial p / \partial x_i = -x_i p$, we obtain

$$
\begin{aligned}
\int_{\mathbb{R}^n} \langle \nabla F, u \rangle p \, dx &= \sum_{i=1}^{n} \int_{\mathbb{R}^n} \frac{\partial F}{\partial x_i} u_i p \, dx \\
&= \sum_{i=1}^{n} \left(- \int_{\mathbb{R}^n} F \frac{\partial u_i}{\partial x_i} p \, dx + \int_{\mathbb{R}^n} F u_i x_i p \, dx \right) \\
&= \int_{\mathbb{R}^n} F \delta(u) p \, dx.
\end{aligned}
$$

This completes the proof. □

3.2 Malliavin Derivative

Let $B = (B_t)_{t \geq 0}$ be a Brownian motion on a probability space $(\Omega, \mathcal{F}, \mathrm{P})$ such that $\mathcal{F}$ is the σ-field generated by B. Set $H = L^2(\mathbb{R}_+)$, and for any $h \in H$, consider the Wiener integral

$$
B(h) = \int_0^\infty h(t) \, dB_t.
$$

The Hilbert space H plays a basic role in the definition of the derivative operator. In fact, the derivative of a random variable $F \colon \Omega \to \mathbb{R}$ takes values in H, and $(D_t F)_{t \geq 0}$ is a stochastic process in $L^2(\Omega; H)$.

We start by defining the derivative in a dense subset of $L^2(\Omega)$. More precisely, consider the set $\mathcal{S}$ of smooth and cylindrical random variables of the form

$$
F = f(B(h_1), \ldots, B(h_n)), \tag{3.1}
$$

where $f \in C_p^\infty(\mathbb{R}^n)$ and $h_i \in H$.

Definition 3.2.1 If $F \in \mathcal{S}$ is a smooth and cylindrical random variable of the form (3.1), the *derivative DF* is the H-valued random variable defined by

$$
D_t F = \sum_{i=1}^{n} \frac{\partial f}{\partial x_i}(B(h_1), \ldots, B(h_n)) h_i(t).
$$

For instance, $D(B(h)) = h$ and $D(B_{t_1}) = \mathbf{1}_{[0, t_1]}$, for any $t_1 \geq 0$.

This defines a linear and unbounded operator from $\mathcal{S} \subset L^2(\Omega)$ into $L^2(\Omega; H)$. Let us now introduce the divergence operator. Denote by $\mathcal{S}_H$ the class of smooth and cylindrical stochastic processes $u = (u_t)_{t \geq 0}$ of the form

$$u_t = \sum_{j=1}^{n} F_j h_j(t), \qquad (3.2)$$

where $F_j \in S$ and $h_j \in H$.

Definition 3.2.2   We define the *divergence* of an element u of the form (3.2) as the random variable given by

$$\delta(u) = \sum_{j=1}^{n} F_j B(h_j) - \sum_{j=1}^{n} \langle DF_j, h_j \rangle_H.$$

In particular, for any $h \in H$ we have $\delta(h) = B(h)$.

As in the finite-dimensional case, the divergence is the adjoint of the derivative operator, as is shown in the next proposition.

Proposition 3.2.3   *Let $F \in S$ and $u \in S_H$. Then*

$$E(F\delta(u)) = E(\langle DF, u \rangle_H).$$

Proof   We can assume that $F = f(B(h_1) \ldots, B(h_n))$ and

$$u = \sum_{j=1}^{n} g_j(B(h_1) \ldots, B(h_n))h_j,$$

where $h_1, \ldots, h_n$ are orthonormal elements in H. In this case, the duality relationship reduces to the finite-dimensional case proved in Proposition 3.1.1.   □

We will make use of the notation $D_h F = \langle DF, h \rangle_H$ for any $h \in H$ and $F \in S$. The following proposition states the basic properties of the derivative and divergence operators on smooth and cylindrical random variables.

Proposition 3.2.4   *Suppose that $u, v \in S_H$, $F \in S$, and $h \in H$. Then, if $(e_i)_{i \geq 1}$ is a complete orthonormal system in H, we have*

$$E(\delta(u)\delta(v)) = E(\langle u, v \rangle_H) + E\left(\sum_{i,j=1}^{\infty} D_{e_i}\langle u, e_j \rangle_H D_{e_j}\langle v, e_i \rangle_H \right), \qquad (3.3)$$

$$D_h(\delta(u)) = \delta(D_h u) + \langle h, u \rangle_H, \qquad (3.4)$$

$$\delta(Fu) = F\delta(u) - \langle DF, u \rangle_H. \qquad (3.5)$$

Property (3.3) can also be written as

$$E(\delta(u)\delta(v)) = E\left(\int_0^{\infty} u_t v_t dt \right) + E\left(\int_0^{\infty} \int_0^{\infty} D_s u_t D_t v_s ds dt \right).$$

Proof of Proposition 3.2.4 We first show property (3.4). Consider $u = \sum_{j=1}^{n} F_j h_j$, where $F_j \in S$ and $h_j \in H$ for $j = 1, \ldots, n$. Then, using $D_h(B(h_j)) = \langle h, h_j \rangle_H$, we obtain

$$D_h(\delta(u)) = D_h\left(\sum_{j=1}^{n} F_j B(h_j) - \sum_{j=1}^{n} \langle DF_j, h_j \rangle_H \right)$$

$$= \sum_{j=1}^{n} F_j \langle h, h_j \rangle_H + \sum_{j=1}^{n} (D_h F_j B(h_j) - \langle D_h(DF_j), h_j \rangle_H)$$

$$= \langle u, h \rangle_H + \delta(D_h u).$$

To show property (3.3), using the duality formula (Proposition 3.2.3) and property (3.4), we get

$$E(\delta(u)\delta(v)) = E(\langle v, D(\delta(u)) \rangle_H)$$

$$= E\left(\sum_{i=1}^{\infty} \langle v, e_i \rangle_H D_{e_i}(\delta(u)) \right)$$

$$= E\left(\sum_{i=1}^{\infty} \langle v, e_i \rangle_H \left(\langle u, e_i \rangle_H + \delta(D_{e_i} u) \right) \right)$$

$$= E(\langle u, v \rangle_H) + E\left(\sum_{i,j=1}^{\infty} D_{e_i}\langle u, e_j \rangle_H \, D_{e_j}\langle v, e_i \rangle_H \right).$$

Finally, to prove property (3.5) we choose a smooth random variable $G \in S$ and write, using the duality relationship (Proposition 3.2.3),

$$E(\delta(Fu)G) = E(\langle DG, Fu \rangle_H) = E(\langle u, D(FG) - GDF \rangle_H)$$

$$= E((\delta(u)F - \langle u, DF \rangle_H)G),$$

which implies the result because S is dense in $L^2(\Omega)$. $\square$

3.3 Sobolev Spaces

The next proposition will play a basic role in extending the derivative to suitable Sobolev spaces of random variables.

Proposition 3.3.1 *The operator D is closable from $L^p(\Omega)$ to $L^p(\Omega; H)$ for any $p \geq 1$.*

Proof Assume that the sequence $F_N \in S$ satisfies

$$F_N \xrightarrow{L^p(\Omega)} 0 \quad \text{and} \quad DF_N \xrightarrow{L^p(\Omega;H)} \eta,$$

as $N \to \infty$. Then $\eta = 0$. Indeed, for any $u = \sum_{j=1}^{N} G_j h_j \in S_H$ such that $G_j B(h_j)$ and DG_j are bounded, by the duality formula (Proposition 3.2.3), we obtain

$$
\begin{aligned}
E(\langle \eta, u \rangle_H) &= \lim_{N \to \infty} E(\langle DF_N, u \rangle_H) \\
&= \lim_{N \to \infty} E(F_N \delta(u)) = 0.
\end{aligned}
$$

This implies that $\eta = 0$, since the set of $u \in S_H$ with the above properties is dense in $L^p(\Omega; H)$ for all $p \geq 1$. $\qquad \square$

We consider the closed extension of the derivative, which we also denote by D. The domain of this operator is defined by the following Sobolev spaces. For any $p \geq 1$, we denote by $\mathbb{D}^{1,p}$ the closure of S with respect to the seminorm

$$
\|F\|_{1,p} = \left(E(|F|^p) + E\left(\left| \int_0^\infty (D_t F)^2 dt \right|^{p/2} \right) \right)^{1/p}.
$$

In particular, F belongs to $\mathbb{D}^{1,p}$ if and only if there exists a sequence $F_n \in S$ such that

$$
F_n \xrightarrow{L^p(\Omega)} F \quad \text{and} \quad DF_n \xrightarrow{L^p(\Omega;H)} DF,
$$

as $n \to \infty$. For $p = 2$, the space $\mathbb{D}^{1,2}$ is a Hilbert space with scalar product

$$
\langle F, G \rangle_{1,2} = E(FG) + E\left(\int_0^\infty D_t F D_t G \, dt \right).
$$

In the same way we can introduce spaces $\mathbb{D}^{1,p}(H)$ by taking the closure of S_H. The corresponding seminorm is denoted by $\| \cdot \|_{1,p,H}$.

The Malliavin derivative satisfies the following chain rule.

Proposition 3.3.2 *Let $\varphi \colon \mathbb{R} \to \mathbb{R}$ be a continuous differentiable function such that $|\varphi'(x)| \leq C(1 + |x|^\alpha)$ for some $\alpha \geq 0$. Let $F \in \mathbb{D}^{1,p}$ for some $p \geq \alpha + 1$. Then, $\varphi(F)$ belongs to $\mathbb{D}^{1,q}$, where $q = p/(\alpha + 1)$, and*

$$
D(\varphi(F)) = \varphi'(F) DF.
$$

Proof Notice that $|\varphi(x)| \leq C'(1 + |x|^{\alpha+1})$, for some constant C', which implies that $\varphi(F) \in L^q(\Omega)$ and, by Hölder's inequality, $\varphi'(F)DF \in L^q(\Omega; H)$. Then, to show the proposition it suffices to approximate F by smooth and cylindrical random variables, and φ by $\varphi * \alpha_n$, where α_n is an approximation of the identity. $\qquad \square$

The chain rule can be extended to the case of Lipschitz functions (see Exercise 3.3).

We next define the domain of the divergence operator. We identify the Hilbert space $L^2(\Omega; H)$ with $L^2(\Omega \times \mathbb{R}_+)$.

Definition 3.3.3 The *domain* of the divergence operator Dom δ in $L^2(\Omega)$ is the set of processes $u \in L^2(\Omega \times \mathbb{R}_+)$ such that there exists $\delta(u) \in L^2(\Omega)$ satisfying the duality relationship

$$E(\langle DF, u \rangle_H) = E(\delta(u)F),$$

for any $F \in \mathbb{D}^{1,2}$.

Observe that δ is a linear operator such that $E(\delta(u)) = 0$. Moreover, δ is closed; that is, if the sequence $u_n \in \mathcal{S}_H$ satisfies

$$u_n \overset{L^2(\Omega;H)}{\longrightarrow} u \quad \text{and} \quad \delta(u_n) \overset{L^2(\Omega)}{\longrightarrow} G,$$

as $n \to \infty$, then u belongs to Dom δ and $\delta(u) = G$.

Proposition 3.2.4 can be extended to random variables in suitable Sobolev spaces. Property (3.3) holds for $u, v \in \mathbb{D}^{1,2}(H) \subset \text{Dom}\,\delta$ (see Exercise 3.5) and, in this case, for any $u \in \mathbb{D}^{1,2}(H)$ we can write

$$E(\delta(u)^2) \leq E\left(\int_0^\infty (u_t)^2 dt \right) + E\left(\int_0^\infty \int_0^\infty (D_s u_t)^2 ds dt \right) = \|u\|_{1,2,H}^2.$$

Property (3.4) holds if $u \in \mathbb{D}^{1,2}(H)$ and $D_h u \in \text{Dom}\,\delta$ (see Exercise 3.6). Finally, property (3.5) holds if $F \in \mathbb{D}^{1,2}$, $Fu \in L^2(\Omega; H)$, $u \in \text{Dom}\,\delta$, and the right-hand side is square integrable (see Exercise 3.7).

We can also introduce iterated derivatives and the corresponding Sobolev spaces. The kth derivative $D^k F$ of a random variable $F \in S$ is the k-parameter process obtained by iteration:

$$D^k_{t_1,\dots,t_k} F = \sum_{i_1,\dots,i_k=1}^n \frac{\partial^k f}{\partial x_{i_1} \cdots \partial x_{i_k}}(B(h_1),\dots,B(h_n))h_{i_1}(t_1)\cdots h_{i_k}(t_k).$$

For any $p \geq 1$, the operator D^k is closable from $L^p(\Omega)$ into $L^p(\Omega; H^{\otimes k})$ (see Exercise 3.8), and we denote by $\mathbb{D}^{k,p}$ the closure of S with respect to the seminorm

$$\|F\|_{k,p} = \left(E(|F|^p) + E\left(\sum_{j=1}^k \left| \int_{\mathbb{R}_+^j} (D^j_{t_1,\dots,t_j}F)^2 dt_1 \cdots dt_j \right|^{p/2} \right) \right)^{1/p}.$$

For any $k \geq 1$, we set $\mathbb{D}^{k,\infty} := \cap_{p \geq 2}\mathbb{D}^{k,p}$, $\mathbb{D}^{\infty,2} := \cap_{k \geq 1}\mathbb{D}^{k,2}$, and $\mathbb{D}^\infty := \cap_{k \geq 1}\mathbb{D}^{k,\infty}$. Similarly, we can introduce the spaces $\mathbb{D}^{k,p}(H)$.

These spaces satisfy that, for any $q \geq p \geq 2$ and $\ell \geq k$, $\mathbb{D}^{\ell,q} \subset \mathbb{D}^{k,p}$. We also have the following Hölder's inequality for the $\| \cdot \|_{k,p}$-seminorms.

Proposition 3.3.4 *Let $p, q, r \geq 2$ such that $1/p + 1/q = 1/r$. Let $F \in \mathbb{D}^{k,p}$ and $G \in \mathbb{D}^{k,q}$. Then FG belongs to $\mathbb{D}^{k,r}$ and*

$$\|FG\|_{k,r} \leq c_{p,q,r} \|F\|_{k,p} \|G\|_{k,q}.$$

Proof The result follows from the Leibnitz rule (see Exercise 3.9) and the usual Hölder's inequality. □

3.4 The Divergence as a Stochastic Integral

The Malliavin derivative is a local operator in the following sense. Let $[a, b] \subset \mathbb{R}_+$ be fixed. We denote by $\mathcal{F}_{[a,b]}$ the σ-field generated by the random variables $\{B_s - B_a, s \in [a, b]\}$.

Lemma 3.4.1 *Let F be a random variable in $\mathbb{D}^{1,2} \cap L^2(\Omega, \mathcal{F}_{[a,b]}, P)$. Then $D_t F = 0$ for almost all $(t, \omega) \in [a, b]^c \times \Omega$.*

Proof If F belongs to $\mathcal{S} \cap L^2(\Omega, \mathcal{F}_{[a,b]}, P)$ then this property is clear. The general case follows by approximation. □

The following result, proved by Gaveau and Trauber (1982), says that the divergence operator is an extension of Itô's integral.

Theorem 3.4.2 *We have $L^2(\mathcal{P}) \subset \mathrm{Dom}\, \delta$ and, for any $u \in L^2(\mathcal{P})$, $\delta(u)$ coincides with Itô's stochastic integral*

$$\delta(u) = \int_0^\infty u_t \, dB_t.$$

Proof Consider a simple process u of the form

$$u_t = \sum_{j=0}^{n-1} \phi_j \mathbf{1}_{(t_j, t_{j+1}]}(t),$$

where $0 \leq t_0 < t_1 < \cdots < t_n$ and the random variables $\phi_j \in \mathcal{S}$ are $\mathcal{F}_{t_j}$-measurable. Then $\delta(u)$ coincides with the Itô integral of u because, by (3.5),

$$\delta(u) = \sum_{j=0}^{n-1} \phi_j (B_{t_{j+1}} - B_{t_j}) - \sum_{j=0}^{n-1} \int_{t_j}^{t_{j+1}} D_t \phi_j \, dt = \sum_{j=0}^{n-1} \phi_j (B_{t_{j+1}} - B_{t_j}),$$

taking into account that $D_t \phi_j = 0$ if $t > t_j$ by Lemma 3.4.1. Then the result follows by approximating any process in $L^2(\mathcal{P})$ by simple processes (see Proposition 2.1.4), and approximating any $\phi_j \in L^2(\Omega, \mathcal{F}_{t_j}, P)$ by $\mathcal{F}_{t_j}$-measurable smooth and cylindrical random variables. □

If u is not adapted, $\delta(u)$ coincides with an anticipating stochastic integral introduced by Skorohod (1975). Using techniques of Malliavin calculus, Nualart and Pardoux (1988) developed a stochastic calculus for the Skorohod integral.

If u and v are adapted then, for $s < t$, $D_t v_s = 0$ and, for $s > t$, $D_s u_t = 0$. As a consequence, property (3.3) leads to the isometry property of Itô's integral for adapted processes $u, v \in \mathbb{D}^{1,2}(H)$:

$$E(\delta(u)\delta(v)) = E\left(\int_0^\infty u_t v_t dt \right).$$

If u is an adapted process in $\mathbb{D}^{1,2}(H)$ then, from property (3.4), we obtain

$$D_t\left(\int_0^\infty u_s dB_s \right) = u_t + \int_t^\infty D_t u_s dB_s, \tag{3.6}$$

because $D_t u_s = 0$ if $t > s$.

The next result shows that we can differentiate Lebesgue integrals of stochastic processes.

Proposition 3.4.3 *For any $u \in \mathbb{D}^{1,2}(H)$ and $h \in H$, we have $\langle u, h \rangle_H \in \mathbb{D}^{1,2}$ and*

$$D_t \langle u, h \rangle_H = \langle D_t u, h \rangle_H.$$

Proof Let $u \in \mathbb{D}^{1,2}(H)$. Then there exists a sequence $u_n \in S_H$ that converges to u in $L^2(\Omega; H)$ and is such that the sequence Du_n converges to Du in $L^2(\Omega; H \otimes H)$. Now, let $h \in H$. Then the sequence $\langle u_n, h \rangle_H$ converges to $\langle u, h \rangle_H$ in $L^2(\Omega)$, and the sequence $D\langle u_n, h \rangle_H$ converges to $D\langle u, h \rangle_H$ in $L^2(\Omega; H)$, which concludes the proof. □

Example 3.4.4 Taking $h = \mathbf{1}_{[0,T]}$, we get

$$D_t \int_0^T u_s \, ds = \int_0^T D_t u_s \, ds.$$

For example,

$$D_t \int_0^T B_s \, ds = \int_0^T D_t B_s \, ds = T - t.$$

3.5 Isonormal Gaussian Processes

So far, we have developed the Malliavin calculus with respect to Brownian motion. In this case, the Wiener integral $B(h) = \int_0^\infty h(t)dB_t$ gives rise to a centered Gaussian family indexed by the Hilbert space $H = L^2(\mathbb{R}_+)$.

More generally, consider a separable Hilbert space H with scalar product $\langle \cdot, \cdot \rangle_H$. An *isonormal Gaussian process* is a centered Gaussian family $\mathcal{H}_1 = \{W(h), h \in H\}$ satisfying

$$E(W(h)W(g)) = \langle h, g \rangle_H,$$

for any $h, g \in H$. Observe that $\mathcal{H}_1$ is a Gaussian subspace of $L^2(\Omega)$. The notion of isonormal Gaussian process was introduced by Segal (1954).

The Malliavin calculus can be developed in the framework of an isonormal Gaussian process, and all the notions and properties that do not depend on the fact that $H = L^2(\mathbb{R}_+)$ can be extended to this more general context. For a complete exposition of the Malliavin calculus with respect to a general isonormal Gaussian process we refer to Nualart (2006). We next give several examples of isonormal Gaussian processes.

Example 3.5.1   Let $(T, \mathcal{B}, \mu)$ be a measure space, where μ is a σ-finite measure. Consider a centered Gaussian family of random variables

$$W = \{W(A), A \in \mathcal{B}, \mu(A) < +\infty\},$$

with covariance

$$E(W(A) \cap W(B)) = \mu(A \cap B).$$

Then W is called a *white noise* on $(T, \mathcal{B}, \mu)$; this is a generalization of the white noise in $D \subset \mathbb{R}^m$ (see Definition 1.3.1). The mapping $\mathbf{1}_A \to W(A)$ can be extended to a linear isometry from $L^2(T)$ to the Gaussian space spanned by W:

$$\varphi \to \int_T \varphi(x) W(dx).$$

Example 3.5.2   Let $X = (X_t)_{t \geq 0}$ be a continuous centered Gaussian process with covariance function

$$R(s, t) = E(X_t X_s).$$

The Gaussian subspace generated by X can be identified with an isonormal Gaussian process $\mathcal{H}_1 = \{X(h), h \in H\}$, where the separable Hilbert space H is defined as follows. We denote by $\mathcal{E}$ the set of step functions on $[0, T]$. Let H be the Hilbert space defined as the closure of $\mathcal{E}$ with respect to the scalar product

$$\langle \mathbf{1}_{[0,t]}, \mathbf{1}_{[0,s]} \rangle_H = R(t, s).$$

The mapping $\mathbf{1}_{[0,t]} \mapsto X_t$ can be extended to an isometry $h \to X(h)$ between H and $\mathcal{H}_1$.

Example 3.5.3 A particular case of Example 3.5.2 is the *fractional Brownian motion*, with Hurst parameter $H \in (0,1)$, denoted $B^H = (B_t^H)_{t \geq 0}$. By definition, B^H is a centered Gaussian process with covariance function given by

$$R_H(t,s) := E(B_t^H B_s^H) = 2^{-1}(t^{2H} + s^{2H} - |t-s|^{2H}). \tag{3.7}$$

This process was introduced by Kolmogorov (1940) and was studied by Mandelbrot and Van Ness (1968), where a stochastic integral representation in terms of a two-sided Brownian motion on the whole real line was established.

One can find a square integrable kernel K_H, whose precise expression is given below, such that

$$\int_0^{t \wedge s} K_H(t,u) K_H(s,u) du = R_H(t,s).$$

This implies the existence of the fractional Brownian through the integral representation $B_t^H = \int_0^t K_H(t,s) dW_s$, where $W = (W_t)_{t \geq 0}$ is a Brownian motion. Conversely, given B^H, we will show that there exists a Brownian motion W such that this integral representation holds.

Note that, for $H = \frac{1}{2}$, B^H is a standard Brownian motion. However, for $H \neq \frac{1}{2}$ the increments are non-independent and are positively correlated for $H > \frac{1}{2}$ and negatively correlated for $H < \frac{1}{2}$.

The form of the covariance function entails that

$$E(|B_t^H - B_s^H|^2) = |t-s|^{2H}.$$

This implies that the process B^H has stationary increments. Moreover, by Kolmogorov's continuity criterion (Theorem A.4.1), the trajectories of B^H are γ-Hölder continuous on $[0,T]$ for any $\gamma < H$ and $T > 0$. Moreover, the process is self-similar in the sense that for any $a > 0$, the processes $(B_t^H)_{t \geq 0}$ and $(a^{-H} B_{at}^H)_{t \geq 0}$ have the same distribution.

The Gaussian subspace generated by B^H can be identified with the isonormal Gaussian process defined in Example 3.5.2. However, we do not know whether the elements of the Hilbert space H introduced in Example 3.5.2 can be considered as real-valued functions. This turn out to be true for $H < \frac{1}{2}$ but false when $H > \frac{1}{2}$, as is explained below.

In the case $H > \frac{1}{2}$, one can show that the kernel $K_H(t,s)$ is given by, for $s < t$,

$$K_H(t,s) = c_H s^{1/2-H} \int_s^t r^{H-3/2}(r-s)^{H-1/2} dr,$$

where c_H is defined as

$$c_H = \left(\frac{H(2H-1)}{\beta(2-2H, H-\frac{1}{2})}\right)^{1/2}.$$

Then, the elements of H may be not functions but distributions of negative order; see Pipiras and Taqqu (2000). One can show that

$$L^2([0,T]) \subset L^{1/H}([0,T]) \subset |H| \subset H,$$

where $|H|$ is the Banach space of measurable functions φ on $[0,T]$ such that

$$\|\varphi\|_{|H|}^2 = H(2H-1) \int_0^T \int_0^T |r-u|^{2H-2} |\varphi(r)| |\varphi(u)| du dr < \infty.$$

In the case where $H < \frac{1}{2}$, the kernel $K_H(t,s)$ is given by, for $s < t$,

$$K_H(t,s) = c_H \left\{ \left(\frac{t}{s}\right)^{H-1/2} (t-s)^{H-1/2} - \left(\frac{1}{2}-H\right) s^{1/2-H} \int_s^t r^{H-3/2} (r-s)^{H-1/2} dr \right\},$$

where

$$c_H = \left(\frac{2H}{(1-2H)\beta(1-2H, H+\frac{1}{2})}\right)^{1/2}.$$

Then, for all $\alpha > \frac{1}{2} - H$, we have that

$$C^\alpha([0,T]) \subset H \subset L^2([0,T]).$$

Consider the operator K_H^* from $\mathcal{E}$ to $L^2([0,T])$ defined as

$$K_H^* \mathbf{1}_{[0,t]}(s) = K_H(t,s) \mathbf{1}_{[0,t]}(s).$$

Then K_H^* is a linear isometry between $\mathcal{E}$ and $L^2([0,T])$ that can be extended to the Hilbert space H.

In the case $H > \frac{1}{2}$ the operator K_H^* can be expressed in terms of fractional integrals, while in the case $H > \frac{1}{2}$ it can be expressed in terms of fractional derivatives. In both cases, one can show that, for any $a \in [0,T]$, the indicator function $\mathbf{1}_{[0,a]}$ belongs to the image of K_H^* and thus $\text{Im}(K_H^*) = L^2([0,T])$. Now, consider the family $W = \{W(\varphi), \varphi \in H\}$ defined as

$$W(\varphi) = B^H((K_H^*)^{-1}\varphi).$$

It is easy to show that W is a family of centered Gaussian random variables with covariance given by

$$E(W(\varphi)W(\psi)) = \langle \varphi, \psi \rangle_{L^2([0,T])} \quad \text{for all } \varphi, \psi \in L^2([0,T]).$$

In particular, the process $W_t = B^H((K_H^*)^{-1}\mathbf{1}_{[0,t]})$ is a Brownian motion.

Moreover, for any $\varphi \in H$, the stochastic integral $B^H(\varphi)$ admits the following representation as a Wiener integral:

$$B^H(\varphi) = \int_0^T K_H^* \varphi(s) dW_s.$$

In particular,

$$B_t^H = \int_0^t K_H(t, s) dW_s.$$

Exercises

3.1 Consider the family $\mathcal{P}$ of random variables of the form $p(B(h_1) \ldots, B(h_n))$, where $h_i \in H$ and p is a polynomial. Show that $\mathcal{P}$ is dense in $L^q(\Omega)$ for all $q \geq 1$.

Hint: Assume that $q > 1$ and let r be the conjugate of q. Show that if $Z \in L^r(\Omega)$ satisfies $E(ZY) = 0$ for all $Y \in \mathcal{P}$ then $Z = 0$.

3.2 Show that Definition 3.2.1 does not depend on the choice of the representation for F.

3.3 Let $\varphi : \mathbb{R} \to \mathbb{R}$ be a function such that

$$|\varphi(x) - \varphi(y)| \leq K|x - y|,$$

for all $x, y \in \mathbb{R}$. Let F be a random variable in the space $\mathbb{D}^{1,2}$. Show that $\varphi(F)$ belongs to $\mathbb{D}^{1,2}$ and that there exists a random variable G bounded by K such that

$$D(\varphi(F)) = GDF.$$

Moreover, show that when the law of the random variable F is absolutely continuous with respect to the Lebesgue measure, then $G = \varphi'(F)$.

Hint: Approximate φ by the sequence $\varphi_n = \varphi * \alpha_n$, where α_n is an approximation of the identity, apply the chain rule in Proposition 3.3.2 to the random variable F and the function φ_n, and use Corollary 4.2.5 to conclude.

3.4 Let $X = (X_t)_{t \in [0,1]}$ be a continuous centered Gaussian process. Assume that a.s. X attains its maximum on a unique random point T at $[0, 1]$. Show that the random variable $M = \sup_{t \in [0,1]} X_t$ belongs to the space $\mathbb{D}^{1,2}$ and that $D_t M = \mathbf{1}_{[0,T]}(t)$.

Hint: Approximate the supremum of X by the maximum on a finite set, and use the chain rule for Lipschitz functions (Exercise 3.3), and Corollary 4.2.5.

3.5 Show that, for any $u, v \in \mathbb{D}^{1,2}(H)$,

$$E(\delta(u)\delta(v)) = E\left(\int_0^\infty u_t v_t dt\right) + E\left(\int_0^\infty \int_0^\infty D_s u_t D_t v_s ds dt\right).$$

3.6 Suppose that $u \in \mathbb{D}^{2,2}(H)$ and let $h \in H$. Show that $D_h u$ belongs to the domain of the divergence, $\delta(u)$ belongs to $\mathbb{D}^{1,2}$, and

$$D_h(\delta(u)) = \delta(D_h u) + \langle h, u \rangle_H.$$

3.7 Show that, for any $F \in \mathbb{D}^{1,2}$ and $u \in \mathrm{Dom}\,\delta$ such that $Fu \in L^2(\Omega; H)$,

$$\delta(Fu) = F\delta(u) - \langle DF, u \rangle_H,$$

provided that the right-hand side is square integrable.

3.8 Show that, for any $p \geq 1$, the operator D^k is closable from $L^p(\Omega)$ into $L^p(\Omega; H^{\otimes k})$.

3.9 Prove the following Leibnitz rule for an iterated derivative: for any random variables $F, G \in S$ and any $k \geq 2$,

$$D^k(FG) = \sum_{i=0}^{k} \binom{k}{i} D^i F D^{k-i} G.$$

4

Wiener Chaos

In this chapter we present the Wiener chaos expansion, which provides an orthogonal decomposition of random variables in $L^2(\Omega)$ in terms of multiple stochastic integrals. We then compute the derivative and the divergence operators, introduced in Chapter 3, on the Wiener chaos expansion. This allows us to derive further properties of these two operators.

4.1 Multiple Stochastic Integrals

Recall that $B = (B_t)_{t\geq0}$ is a Brownian motion defined on a probability space $(\Omega, \mathcal{F}, P)$ such that $\mathcal{F}$ is generated by B. Let $L_s^2(\mathbb{R}_+^n)$ be the space of symmetric square integrable functions $f\colon \mathbb{R}_+^n \to \mathbb{R}$. If $f\colon \mathbb{R}_+^n \to \mathbb{R}$, we define its symmetrization by

$$\tilde{f}(t_1,\ldots,t_n) = \frac{1}{n!} \sum_\sigma f(t_{\sigma(1)},\ldots,t_{\sigma(n)}),$$

where the sum runs over all permutations σ of $\{1, 2, \ldots, n\}$. Observe that

$$\|\tilde{f}\|_{L^2(\mathbb{R}_+^n)} \leq \|f\|_{L^2(\mathbb{R}_+^n)}.$$

Definition 4.1.1 The *multiple stochastic integral* of $f \in L_s^2(\mathbb{R}_+^n)$ is defined as the iterated stochastic integral

$$I_n(f) = n! \int_0^\infty \int_0^{t_n} \cdots \int_0^{t_2} f(t_1,\ldots,t_n)dB_{t_1} \cdots dB_{t_n}.$$

Note that if $f \in L^2(\mathbb{R}_+)$, $I_1(f) = B(f)$ is the Wiener integral of f. If $f \in L^2(\mathbb{R}_+^n)$ is not necessarily symmetric, we define

$$I_n(f) = I_n(\tilde{f}).$$

Using the properties of Itô's stochastic integral, one can easily check (see

Exercise 4.1) the following isometry property: for all $n, m \geq 1$, $f \in L^2(\mathbb{R}^n_+)$, and $g \in L^2(\mathbb{R}^m_+)$,

$$E(I_n(f)I_m(g)) = \begin{cases} 0 & \text{if } n \neq m, \\ n!\langle \tilde{f}, \tilde{g}\rangle_{L^2(\mathbb{R}^n_+)} & \text{if } n = m. \end{cases} \tag{4.1}$$

Next, we want to compute the product of two multiple integrals. Let $f \in L^2_s(\mathbb{R}^n_+)$ and $g \in L^2_s(\mathbb{R}^m_+)$. For any $r = 0, \ldots, n \wedge m$, we define the *contraction* of f and g of order r to be the element of $L^2(\mathbb{R}^{n+m-2r}_+)$ defined by

$$(f \otimes_r g)(t_1, \ldots, t_{n-r}, s_1, \ldots, s_{m-r})$$
$$= \int_{\mathbb{R}^r_+} f(t_1, \ldots, t_{n-r}, x_1, \ldots, x_r) g(s_1, \ldots, s_{m-r}, x_1, \ldots, x_r) dx_1 \cdots dx_r.$$

We denote by $f \tilde{\otimes}_r g$ the symmetrization of $f \otimes_r g$. Then, the product of two multiple stochastic integrals satisfies the following formula:

$$I_n(f)I_m(g) = \sum_{r=0}^{n \wedge m} r! \binom{n}{r}\binom{m}{r} I_{n+m-2r}(f \otimes_r g). \tag{4.2}$$

The nth Hermite polynomial is defined by

$$h_n(x) = (-1)^n e^{x^2/2} \frac{d^n}{dx^n}(e^{-x^2/2}). \tag{4.3}$$

The first few Hermite polynomials are $h_0(x) = 1$, $h_1(x) = x$, $h_2(x) = x^2 - 1$, $h_3(x) = x^3 - 3x$, ... It is easy to see (see Exercise 4.3) that these polynomials satisfy the following series expansion:

$$\exp\left(tx - \frac{t^2}{2}\right) = \sum_{n=0}^{\infty} \frac{t^n}{n!} h_n(x). \tag{4.4}$$

The next result gives the relation between multiple stochastic integrals and Hermite polynomials.

Proposition 4.1.2   *For any $g \in L^2(\mathbb{R}_+)$, we have*

$$I_n(g^{\otimes n}) = \|g\|^n_{L^2(\mathbb{R}_+)} h_n\left(\frac{B(g)}{\|g\|_{L^2(\mathbb{R}_+)}}\right),$$

where $g^{\otimes n}(t_1, \ldots, t_n) = g(t_1) \cdots g(t_n)$.

Proof   We can assume that $\|g\|_{L^2(\mathbb{R}_+)} = 1$. We proceed by induction over n. The case $n = 1$ is immediate. We then assume that the result holds for

$1, \ldots, n$. Using the product rule (4.2), the induction hypothesis, and the recursive relation for the Hermite polynomials (4.20), we get

$$I_{n+1}(g^{\otimes(n+1)}) = I_n(g^{\otimes n})I_1(g) - nI_{n-1}(g^{\otimes(n-1)})$$
$$= h_n(B(g))B(g) - nh_{n-1}(B(g))$$
$$= h_{n+1}(B(g)),$$

which concludes the proof. $\square$

The next result is the *Wiener chaos expansion.*

Theorem 4.1.3 *Every $F \in L^2(\Omega)$ can be uniquely expanded into a sum of multiple stochastic integrals as follows:*

$$F = E(F) + \sum_{n=1}^{\infty} I_n(f_n),$$

where $f_n \in L_s^2(\mathbb{R}_+^n)$.

For any $n \geq 1$, we denote by $\mathcal{H}_n$ the closed subspace of $L^2(\Omega)$ formed by all multiple stochastic integrals of order n. For $n = 0$, $\mathcal{H}_0$ is the space of constants. Observe that $\mathcal{H}_1$ coincides with the isonormal Gaussian process $\{B(f), f \in L^2(\mathbb{R}_+)\}$. Then Theorem 4.1.3 can be reformulated by saying that we have the orthogonal decomposition

$$L^2(\Omega) = \oplus_{n=0}^{\infty} \mathcal{H}_n.$$

Proof of Theorem 4.1.3 It suffices to show that if a random variable $G \in L^2(\Omega)$ is orthogonal to $\oplus_{n=0}^{\infty} \mathcal{H}_n$ then $G = 0$. This assumption implies that G is orthogonal to all random variables of the form $B(g)^k$, where $g \in L^2(\mathbb{R}_+)$, $k \geq 0$. This in turn implies that G is orthogonal to all the exponentials $\exp(B(h))$, which form a total set in $L^2(\Omega)$. So $G = 0$. $\square$

4.2 Derivative Operator on the Wiener Chaos

Let us compute the derivative of a multiple stochastic integral.

Proposition 4.2.1 *Let $f \in L_s^2(\mathbb{R}_+^n)$. Then $I_n(f) \in \mathbb{D}^{1,2}$ and*

$$D_t I_n(f) = nI_{n-1}(f(\cdot, t)).$$

Proof Assume that $f = g^{\otimes n}$, with $\|g\|_{L^2(\mathbb{R}_+)} = 1$. Then, using Proposition 4.1.2 and the property (4.19) below of Hermite polynomials, we have

$$D_t I_n(f) = D_t(h_n(B(g))) = h_n'(B(g))D_t(B(g)) = nh_{n-1}(B(g))g(t)$$
$$= ng(t)I_{n-1}(g^{\otimes(n-1)}) = nI_{n-1}(f(\cdot, t)).$$

The general case follows using linear combinations and a density argument. This finishes the proof.   □

Moreover, applying (4.1), we have

$$\mathrm{E}\left(\int_{\mathbb{R}_+}(D_tI_n(f))^2dt\right) = n^2\int_{\mathbb{R}_+}\mathrm{E}(I_{n-1}(f(\cdot,t))^2)dt$$

$$= n^2(n-1)!\int_{\mathbb{R}_+}\|f(\cdot,t)\|^2_{L^2(\mathbb{R}_+^{n-1})}dt$$

$$= nn!\|f\|^2_{L^2(\mathbb{R}_+^n)}$$

$$= n\mathrm{E}(I_n(f)^2). \tag{4.5}$$

As a consequence of Proposition 4.2.1 and (4.5), we deduce the following result.

Proposition 4.2.2 *Let* $F \in L^2(\Omega)$ *with Wiener chaos expansion* $F = \sum_{n=0}^{\infty}I_n(f_n)$. *Then* $F \in \mathbb{D}^{1,2}$ *if and only if*

$$\mathrm{E}(\|DF\|^2_H) = \sum_{n=1}^{\infty}nn!\|f_n\|^2_{L^2(\mathbb{R}_+^n)} < \infty,$$

and in this case

$$D_tF = \sum_{n=1}^{\infty}nI_{n-1}(f_n(\cdot,t)).$$

Similarly, if $k \geq 2$, one can show that $F \in \mathbb{D}^{k,2}$ if and only if

$$\sum_{n=1}^{\infty}n^kn!\|f_n\|^2_{L^2(\mathbb{R}_+^n)} < \infty,$$

and in this case

$$D^k_{t_1,\ldots,t_k}F = \sum_{n=k}^{\infty}n(n-1)\cdots(n-k+1)I_{n-k}(f_n(\cdot,t_1,\ldots,t_k)),$$

where the series converges in $L^2(\Omega \times \mathbb{R}_+^k)$. As a consequence, if $F \in \mathbb{D}^{\infty,2}$ then the following formula, due to Stroock (1987) (see Exercise 4.7), allows us to compute explicitly the kernels in the Wiener chaos expansion of F:

$$f_n = \frac{1}{n!}\mathrm{E}(D^nF). \tag{4.6}$$

Example 4.2.3 Consider $F = B_1^3$. Then

$$f_1(t_1) = E(D_{t_1} B_1^3) = 3E(B_1^2)1_{[0,1]}(t_1) = 31_{[0,1]}(t_1),$$
$$f_2(t_1, t_2) = \tfrac{1}{2}E(D_{t_1,t_2}^2 B_1^3) = 3E(B_1)1_{[0,1]}(t_1 \vee t_2) = 0,$$
$$f_3(t_1, t_2, t_3) = \tfrac{1}{6}E(D_{t_1,t_2,t_3}^3 B_1^3) = 1_{[0,1]}(t_1 \vee t_2 \vee t_3),$$

and we obtain the Wiener chaos expansion

$$B_1^3 = 3B_1 + 6 \int_0^1 \int_0^{t_1} \int_0^{t_2} dB_{t_1} dB_{t_2} dB_{t_3}.$$

Proposition 4.2.2 implies the following characterization of the space $\mathbb{D}^{1,2}$.

Proposition 4.2.4 *Let $F \in L^2(\Omega)$. Assume that there exists an element $u \in L^2(\Omega; H)$ such that, for all $G \in S$ and $h \in H$, the following duality formula holds:*

$$E(\langle u, h \rangle_H G) = E(F\delta(Gh)). \tag{4.7}$$

Then $F \in \mathbb{D}^{1,2}$ and $DF = u$.

Proof Let $F = \sum_{n=0}^{\infty} I_n(f_n)$, where $f_n \in L_s^2(\mathbb{R}_+^n)$. By the duality formula (Proposition 3.2.3) and Proposition 4.2.1, we obtain

$$E(F\delta(Gh)) = \sum_{n=0}^{\infty} E(I_n(f_n)\delta(Gh)) = \sum_{n=0}^{\infty} E(\langle D(I_n(f_n)), h \rangle_H G)$$

$$= \sum_{n=1}^{\infty} E(\langle n I_{n-1}(f_n(\cdot, t)), h \rangle_H G).$$

Then, by (4.7), we get

$$\sum_{n=1}^{\infty} E(\langle n I_{n-1}(f_n(\cdot, t)), h \rangle_H G) = E(\langle u, h \rangle_H G),$$

which implies that the series $\sum_{n=1}^{\infty} n I_{n-1}(f_n(\cdot, t))$ converges in $L^2(\Omega; H)$ and its sum coincides with u. Proposition 4.2.2 allows us to conclude the proof. $\square$

Corollary 4.2.5 *Let $(F_n)_{n \geq 1}$ be a sequence of random variables in $\mathbb{D}^{1,2}$ that converges to F in $L^2(\Omega)$ and is such that*

$$\sup_n E(\|DF_n\|_H^2) < \infty.$$

Then F belongs to $\mathbb{D}^{1,2}$ and the sequence of derivatives $(DF_n)_{n \geq 1}$ converges to DF in the weak topology of $L^2(\Omega; H)$.

Proof The assumptions imply that there exists a subsequence $(F_{n(k)})_{k\geq1}$ such that the sequence of derivatives $(DF_{n(k)})_{k\geq1}$ converges in the weak topology of $L^2(\Omega; H)$ to some element $\alpha \in L^2(\Omega; H)$. By Proposition 4.2.4, it suffices to show that, for all $G \in S$ and $h \in H$,

$$E(\langle\alpha, h\rangle_H G) = E(F\delta(Gh)). \tag{4.8}$$

By the duality formula (Proposition 3.2.3), we have

$$E(\langle DF_{n(k)}, h\rangle_H G) = E(F_{n(k)}\delta(Gh)).$$

Then, taking the limit as k tends to infinity, we obtain (4.8), which concludes the proof.   □

The next proposition shows that the indicator function of a set $A \in \mathcal{F}$ such that $0 < P(A) < 1$ does not belong to $\mathbb{D}^{1,2}$.

Proposition 4.2.6 *Let $A \in \mathcal{F}$ and suppose that the indicator function of A belongs to the space $\mathbb{D}^{1,2}$. Then, $P(A)$ is zero or one.*

Proof Consider a continuously differentiable function φ with compact support, such that $\varphi(x) = x^2$ for each $x \in [0, 1]$. Then, by Proposition 3.3.2, we can write

$$D\mathbf{1}_A = D[(\mathbf{1}_A)^2] = D[\varphi(\mathbf{1}_A)] = 2\mathbf{1}_A D\mathbf{1}_A.$$

Therefore $D\mathbf{1}_A = 0$ and, from Proposition 4.2.2, we deduce that $\mathbf{1}_A = P(A)$, which completes the proof.   □

4.3 Divergence on the Wiener Chaos

We now compute the divergence operator on the Wiener chaos expansion. A square integrable stochastic process $u = (u_t)_{t\geq0} \in L^2(\Omega \times \mathbb{R}_+)$ has an orthogonal expansion of the form

$$u_t = \sum_{n=0}^{\infty} I_n(f_n(\cdot, t)),$$

where $f_0(t) = E(u_t)$ and, for each $n \geq 1$, $f_n \in L^2(\mathbb{R}_+^{n+1})$ is a symmetric function in the first n variables.

Proposition 4.3.1 *The process u belongs to the domain of δ if and only if the series*

$$\delta(u) = \sum_{n=0}^{\infty} I_{n+1}(\tilde{f}_n) \tag{4.9}$$

converges in $L^2(\Omega)$.

Proof Suppose that $G = I_n(g)$ is a multiple stochastic integral of order $n \geq 1$, where g is symmetric. Then

$$
\begin{aligned}
\mathrm{E}(\langle u, DG \rangle_H) &= \int_{\mathbb{R}_+} \mathrm{E}\left(I_{n-1}(f_{n-1}(\cdot, t)) n I_{n-1}(g(\cdot, t))\right) dt \\
&= n(n-1)! \int_{\mathbb{R}_+} \langle f_{n-1}(\cdot, t), g(\cdot, t) \rangle_{L^2(\mathbb{R}_+^{n-1})} \, dt \\
&= n! \langle f_{n-1}, g \rangle_{L^2(\mathbb{R}_+^n)} = n! \langle \tilde{f}_{n-1}, g \rangle_{L^2(\mathbb{R}_+^n)} \\
&= \mathrm{E}(I_n(\tilde{f}_{n-1}) I_n(g)) = \mathrm{E}(I_n(\tilde{f}_{n-1}) G).
\end{aligned}
$$

If $u \in \mathrm{Dom}\,\delta$, we deduce that

$$
\mathrm{E}(\delta(u) G) = \mathrm{E}(I_n(\tilde{f}_{n-1}) G)
$$

for every $G \in \mathcal{H}_n$. This implies that $I_n(\tilde{f}_{n-1})$ coincides with the projection of $\delta(u)$ on the nth Wiener chaos. Consequently, the series in (4.9) converges in $L^2(\Omega)$ and its sum is equal to $\delta(u)$. The converse can be proved by similar arguments. $\qquad\square$

4.4 Directional Derivative

Suppose that $(\Omega, \mathcal{F}, \mathrm{P})$ is the Wiener space introduced in Section 1.4. The Malliavin derivative can be interpreted as a directional derivative in the direction of functions in the Cameron–Martin space, defined as

$$
\mathcal{H} = \left\{ g \in \Omega : g(t) = \int_0^t \dot{g}(s) ds \,, \dot{g} \in L^2(\mathbb{R}_+) \right\}.
$$

For any fixed $g \in \mathcal{H}$, we consider the shift transformation $\tau_g : \Omega \to \Omega$ defined by $\tau_g(\omega) = \omega + g$. Notice that, for all $h \in H$,

$$
\tau_g(B(h)) = B(h) + \langle h, \dot{g} \rangle_H.
$$

By Girsanov's theorem (Theorem 2.10.3), there exists a probability Q, equivalent to P, such that the process $B_t + g(t)$ is a Brownian motion under Q. Therefore, if $F \in L^2(\Omega)$, the random variable $F \circ \tau_g(\omega) = F(\omega + g)$ is well defined almost surely.

Definition 4.4.1 We define the *directional derivative* of $F \in L^2(\Omega)$ in the direction $g \in \mathcal{H}$ as the following limit in probability:

$$
\tilde{D}_g F = \lim_{\epsilon \to 0} \frac{1}{\epsilon} (F \circ \tau_{\epsilon g} - F),
$$

if it exists.

We next define the space $\tilde{\mathbb{D}}^{1,2}$ as the set of random variables $F \in L^2(\Omega)$ satisfying:

(1) for any $g \in \mathcal{H}$, there exists a version of the process $(F^g \circ \tau_{\epsilon g})_{\epsilon \geq 0}$ whose trajectories are absolutely continuous with respect to the Lebesgue measure;

(2) there exists a process $u \in L^2(\Omega \times \mathbb{R}_+)$ such that for any $g \in \mathcal{H}$ the directional derivative $\tilde{D}_g F$ exists, and $\tilde{D}_g F = \langle u, \dot{g} \rangle_H$.

The following characterization of the space $\mathbb{D}^{1,2}$ is due to Sugita (1985).

Theorem 4.4.2   *We have* $\mathbb{D}^{1,2} = \tilde{\mathbb{D}}^{1,2}$, *and the Malliavin derivative of* $F \in \tilde{\mathbb{D}}^{1,2}$ *is* $DF = u$, *where u is the process appearing in* (2).

Proof   We first show that $\mathbb{D}^{1,2} \subset \tilde{\mathbb{D}}^{1,2}$. Let $F \in \mathbb{D}^{1,2}$. It suffices to show that F satisfies conditions (1) and (2). We start by proving condition (2). Since $F \in \mathbb{D}^{1,2}$, there exists a sequence F_n in S that converges to F in $L^2(\Omega)$ and for which the sequence DF_n converges to DF in $L^2(\Omega; H)$. We now fix $g \in \mathcal{H}$ and $\epsilon > 0$, and write

$$\mathrm{E}\left(\left|\frac{1}{\epsilon}(F \circ \tau_{\epsilon g} - F) - \langle DF, \dot{g} \rangle_H\right|\right) \leq \mathrm{E}\left(\left|\frac{1}{\epsilon}(F \circ \tau_{\epsilon g} - F) - \frac{1}{\epsilon}(F_n \circ \tau_{\epsilon g} - F_n)\right|\right)$$
$$+ \mathrm{E}\left(\left|\frac{1}{\epsilon}(F_n \circ \tau_{\epsilon g} - F_n) - \langle DF_n, \dot{g} \rangle_H\right|\right)$$
$$+ \mathrm{E}(|\langle DF_n, \dot{g} \rangle_H - \langle DF, \dot{g} \rangle_H|). \tag{4.10}$$

Clearly, the third term on the right-hand side of (4.10) converges to 0 as n tends to infinity. For fixed n, the second term converges to 0 as $\epsilon \to 0$ because $F_n \in S$. Thus, it suffices to show that the first term converges to 0 as n tends to infinity uniformly in $\epsilon \in (0, 1]$. Since $F_n \in S$, it is easy to show that

$$\frac{1}{\epsilon}(F_n \circ \tau_{\epsilon g} - F_n) = \frac{1}{\epsilon} \int_0^\epsilon \langle DF_n \circ \tau_{\alpha g}, \dot{g} \rangle_H d\alpha.$$

By Girsanov's theorem (Theorem 2.10.3), taking the limit in $L^1(\Omega)$ as $n \to \infty$ of each member in the above equality yields

$$\frac{1}{\epsilon}(F \circ \tau_{\epsilon g} - F) = \frac{1}{\epsilon} \int_0^\epsilon \langle DF \circ \tau_{\alpha g}, \dot{g} \rangle_H d\alpha. \tag{4.11}$$

Finally,

$$\sup_{\epsilon\in(0,1]} \mathrm{E}\left(\left|\frac{1}{\epsilon}\int_0^\epsilon \langle DF\circ\tau_{\alpha g},\dot{g}\rangle_H d\alpha - \int_0^\epsilon \langle DF_n\circ\tau_{\alpha g},\dot{g}\rangle_H d\alpha\right|\right)$$
$$\leq \|\dot{g}\|_H \sup_{\alpha\in(0,1]} \mathrm{E}(\|DF_n - DF\|_H \circ \tau_{\alpha g}),$$

which converges to zero as $n\to\infty$ by Girsanov's theorem. We conclude that F satisfies condition (2) with $u = DF$.

Furthermore, condition (2) is an immediate consequence of (4.11).

We now show that $\tilde{\mathbb{D}}^{1,2} \subset \mathbb{D}^{1,2}$. Let $F\in L^2(\Omega)$ and $g\in\mathcal{H}$, and assume that (1) and (2) hold. We need to show that F belongs to $\mathbb{D}^{1,2}$ and that $DF = u$, where u is the process in (2). We divide the proof into steps.

Step 1   Fix $g\in\mathcal{H}$. Condition (1) implies that there exists a version of the process $(F\circ\tau_{\epsilon g})_{\epsilon\geq 0}$ which is differentiable almost everywhere in $\mathbb{R}$. That is, for all $\omega\in\Omega$ and for almost all $\epsilon\in\mathbb{R}_+$,

$$\frac{d}{d\epsilon}(F\circ\tau_{\epsilon g})(\omega) = \lim_{\alpha\to 0}\frac{1}{\alpha}(F\circ\tau_{(\epsilon+\alpha)g}(\omega) - F\circ\tau_{\epsilon g}(\omega)). \tag{4.12}$$

Then, by Fubini's theorem, this implies that there exists a Borel set $N\subset\mathbb{R}$ with $\ell(N) = 0$ such that, for all $\epsilon\notin N$, (4.12) holds for almost all $\omega\in\Omega$.

By Definition 4.4.1, we get the convergence in probability

$$\lim_{\alpha\to 0}\frac{1}{\alpha}(F\circ\tau_{\alpha g} - F) = \tilde{D}_g F.$$

As a consequence, by Girsanov's theorem we obtain the following convergence in probability for any $\epsilon\in\mathbb{R}$

$$\lim_{\alpha\to 0}\frac{1}{\alpha}(F\circ\tau_{(\alpha+\epsilon)g} - F\circ\tau_{\epsilon g}) = (\tilde{D}_g F)\circ\tau_{\epsilon g}. \tag{4.13}$$

Then, from (4.12) and (4.13), we deduce that, for almost all (ϵ,ω),

$$\frac{d}{d\epsilon}(F\circ\tau_{\epsilon g})(\omega) = (\tilde{D}_g F)\circ\tau_{\epsilon g}(\omega). \tag{4.14}$$

Step 2   By Proposition 4.2.4, it suffices to show that there exists $u\in L^2(\Omega\times\mathbb{R}_+)$ such that, for all $G\in S$ and $h\in H$,

$$\mathrm{E}(\langle u,h\rangle_H G) = \mathrm{E}(F\delta(Gh)). \tag{4.15}$$

By (2), (4.15) is equivalent to showing that

$$\mathrm{E}(\tilde{D}_g FG) = \mathrm{E}(F\delta(Gh)), \tag{4.16}$$

where $\dot{g} = h$.

Step 3   Using that an absolutely continuous function is the integral of its derivative, and applying (4.14), we can write

$$\tilde{D}_g F = \lim_{\alpha \to 0} \frac{1}{\alpha}(F \circ \tau_{\alpha g} - F) = \lim_{\alpha \to 0} \frac{1}{\alpha} \int_0^\alpha (\tilde{D}_g F) \circ \tau_{\epsilon g} d\epsilon, \qquad (4.17)$$

where the convergence holds in probability. Applying Girsanov's theorem, one can show that

$$\sup_{\epsilon \in (0,1]} \left\| (\tilde{D}_g F) \circ \tau_{\epsilon g} \right\|_q < \infty,$$

for any $1 \le q < 2$. Therefore, the convergence in (4.17) is in $L^q(\Omega)$ for all $1 \le q < 2$.

Step 4   By Girsanov's theorem,

$$\frac{1}{\alpha}(E((F \circ \tau_{\alpha g})G) - E(FG)) = E\left(F \frac{1}{\alpha}\left((G \circ \tau_{-\alpha g}) \exp\left(\alpha B(h) - \tfrac{1}{2}\alpha^2 \|h\|_H^2\right) - G\right)\right).$$

Moreover, we have, almost surely and in $L^p(\Omega)$ for any $p \ge 1$,

$$\lim_{\alpha \to 0} \frac{1}{\alpha}\left((G \circ \tau_{-\alpha g}) \exp\left(\alpha B(h) - \tfrac{1}{2}\alpha^2 \|h\|_H^2\right) - G\right)$$
$$= \frac{d}{d\alpha}\left((G \circ \tau_{-\alpha g}) \exp\left(\alpha B(h) - \tfrac{1}{2}\alpha^2 \|h\|_H^2\right)\right)\Big|_{\alpha=0}$$
$$= B(h)G - \langle DG, h \rangle_H = \delta(Gh).$$

Taking into account (4.17), this implies that

$$E(\tilde{D}_g FG) = \lim_{\alpha \to 0} \frac{1}{\alpha}E((F \circ \tau_{\alpha g} - F)G) = E(F\delta(Gh)),$$

which concludes the proof of (4.16).   □

Exercises

4.1   Show that, for all $n, m \ge 1$, $f \in L^2(\mathbb{R}_+^n)$, and $g \in L^2(\mathbb{R}_+^m)$,

$$E(I_n(f)I_m(g)) = n! \langle \tilde{f}, \tilde{g} \rangle_{L^2(\mathbb{R}_+^n)} 1_{\{n=m\}}.$$

4.2   Using Itô's formula, show that, for any $f, g \in L_s^2(\mathbb{R}_+^2)$,

$$I_2(f)I_2(g) = \sum_{r=0}^{2} r! \binom{2}{r}\binom{2}{r} I_{4-2r}(f \otimes_r g).$$

4.3   Check that the Hermite polynomials satisfy

$$\exp\left(tx - \frac{t^2}{2}\right) = \sum_{n=0}^{\infty} \frac{t^n}{n!} h_n(x). \qquad (4.18)$$

4.4 Use (4.18) to show the following properties of the Hermite polynomials:

$$h'_n(x) = nh_{n-1}(x), \tag{4.19}$$

$$h_{n+1}(x) = xh_n(x) - nh_{n-1}(x), \tag{4.20}$$

$$h_n(-x) = (-1)^n h_n(x).$$

4.5 For every n we define the extended Hermite polynomial $h_n(x, \lambda)$ by

$$h_n(x, \lambda) = \lambda^{n/2} h_n\left(\frac{x}{\sqrt{\lambda}}\right), \quad \text{for } x \in \mathbb{R} \text{ and } \lambda > 0.$$

Check that

$$\exp\left(tx - \frac{t^2 \lambda}{2}\right) = \sum_{n=0}^{\infty} \frac{t^n}{n!} h_n(x, \lambda).$$

Show that, for any $g \in L^2(\mathbb{R}_+)$,

$$I_n(g^{\otimes n}) = h_n(B(g), \|g\|^2_{L^2(\mathbb{R}_+)}).$$

4.6 Show that the process $(h_n(B_t, t))_{t \geq 0}$ is a martingale.

4.7 Show that if $F \in \mathbb{D}^{\infty,2}$ then the kernels in the Wiener chaos expansion of F satisfy the following formula:

$$f_n = \frac{1}{n!} E(D^n F).$$

4.8 Let $F = \exp\left(B(h) - \frac{1}{2} \int_{\mathbb{R}_+} h^2(s) ds\right)$, $h \in H$. Compute the iterated derivatives of F and the kernels of the Wiener chaos expansion.

4.9 Compute the Wiener chaos expansions of the following random variables:

$$e^{B_T - T/2}, \quad B_T^5, \quad \int_0^1 (t^3 B_t^3 + 2t B_t^2) dB_t, \quad \int_0^1 t e^{B_t} dB_t.$$

4.10 Let $F \in \mathbb{D}^{1,2}$ be a random variable such that $E(F^{-2}) < \infty$. Show that $P(F > 0)$ is zero or one.

Hint: Using the duality relation, compute $E(\text{sign}(F)\delta(u))$, where u is an arbitrary bounded process in the domain of δ, using an approximation of the sign function.

5

Ornstein–Uhlenbeck Semigroup

In this chapter we describe the main properties of the Ornstein–Uhlenbeck semigroup and its generator. We then give the relationship between the Malliavin derivative, the divergence operator, and the Ornstein–Uhlenbeck semigroup generator. This will lead to an important integration-by-parts formula that is crucial in applications. In particular, we present an explicit formula for the density of a centered random variable, which is a consequence of this integration-by-parts formula and can be used to derive upper and lower bounds for the density.

5.1 Mehler's Formula

Let $B = (B_t)_{t \geq 0}$ be a Brownian motion on a probability space $(\Omega, \mathcal{F}, P)$ such that $\mathcal{F}$ is generated by B. Let F be a random variable in $L^2(\Omega)$ with the Wiener chaos decomposition $F = \sum_{n=0}^{\infty} I_n(f_n)$, $f_n \in L_s^2(\mathbb{R}_+^n)$.

Definition 5.1.1 The *Ornstein–Uhlenbeck semigroup* is the one-parameter semigroup $(T_t)_{t \geq 0}$ of operators on $L^2(\Omega)$ defined by

$$T_t(F) = \sum_{n=0}^{\infty} e^{-nt} I_n(f_n).$$

An alternative and useful expression for the Ornstein–Uhlenbeck semigroup is *Mehler's formula*:

Proposition 5.1.2 *Let $B' = (B_t')_{t \geq 0}$ be an independent copy of B. Then, for any $t \geq 0$ and $F \in L^2(\Omega)$, we have*

$$T_t(F) = E'(F(e^{-t}B + \sqrt{1 - e^{-2t}}B')), \tag{5.1}$$

where E' denotes the mathematical expectation with respect to B'.

Proof Both T_t in Definition 5.1.1 and the right-hand side of (5.1) give rise to linear contraction operators on $L^p(\Omega)$, for all $p \geq 1$. For the first

operator, this is clear. For the second, using Jensen's inequality it follows
that, for any $p \geq 1$,

$$E(|T_t(F)|^p) = E(|E'(F(e^{-t}B + \sqrt{1 - e^{-2t}}B'))|^p)$$

$$\leq E(E'(|F(e^{-t}B + \sqrt{1 - e^{-2t}}B')|^p)) = E(|F|^p).$$

Thus, it suffices to show (5.1) for random variables of the form
$F = \exp\left(\lambda B(h) - \frac{1}{2}\lambda^2\right)$, where $B(h) = \int_{\mathbb{R}_+} h_t dB_t$, $h \in H$, is an element
of norm one, and $\lambda \in \mathbb{R}$. We have, using formula (4.4),

$$E'\left(\exp\left(e^{-t}\lambda B(h) + \sqrt{1 - e^{-2t}}\lambda B'(h) - \frac{1}{2}\lambda^2\right)\right)$$

$$= \exp\left(e^{-t}\lambda B(h) - \frac{1}{2}e^{-2t}\lambda^2\right) = \sum_{n=0}^{\infty} e^{-nt}\frac{\lambda^n}{n!}h_n(B(h)) = T_t F,$$

because

$$F = \sum_{n=0}^{\infty} \frac{\lambda^n}{n!}h_n(B(h))$$

and $h_n(B(h)) = I_n(h^{\otimes n})$ (see Proposition 4.1.2). This completes the proof.
$\square$

Mehler's formula implies that the operator T_t is nonnegative. Moreover,
T_t is symmetric, that is,

$$E(GT_t(F)) = E(FT_t(G)) = \sum_{n=0}^{\infty} e^{-nt}E(I_n(f_n)I_n(g_n)),$$

where $F = \sum_{n=0}^{\infty} I_n(f_n)$ and $G = \sum_{n=0}^{\infty} I_n(g_n)$.

The Ornstein–Uhlenbeck semigroup has the following hypercontractivity property, which is due to Nelson (1973). We provide below a proof
of this property using Itô's formula, according to the approach of Neveu
(1976).

Theorem 5.1.3 *Let* $F \in L^p(\Omega)$, $p > 1$, *and* $q(t) = e^{2t}(p - 1) + 1 > p$,
$t > 0$. *Then*

$$\|T_t F\|_{q(t)} \leq \|F\|_p.$$

Proof Let q' be such that $1/q + 1/q' = 1$. It suffices to show that

$$|E((T_t F)G)| \leq \|F\|_p \|G\|_{q'}, \tag{5.2}$$

for any $F \in L^p(\Omega)$ and $G \in L^{q'}(\Omega)$. Because the operator T_t satisfies $|T_t F| \leq
|F|$, we can assume that the random variables F and G are nonnegative. By

an approximation argument, we can also assume that $0 < a \leq F, G \leq b < \infty$. Let B and $\tilde{B}$ be independent Brownian motions and set $\hat{B} = e^{-t}B + \sqrt{1 - e^{-2t}}\tilde{B}$. Then, using Mehler's formula (5.1),

$$E((T_t F)G) = E(\tilde{E}(F(\hat{B}))G) = E(F(\hat{B})G),$$

and inequality (5.2) follows from Exercise 2.15.   □

As a consequence of the hypercontractivity property, for any $1 < p < q < \infty$ the norms $\|\cdot\|_p$ and $\|\cdot\|_q$ are equivalent on any Wiener chaos $\mathcal{H}_n$. In fact, putting $q = e^{2t}(p-1) + 1 > p$ with $t > 0$, we obtain, for every $F \in \mathcal{H}_n$,

$$e^{-nt}\|F\|_q = \|T_t F\|_q \leq \|F\|_p,$$

which implies that

$$\|F\|_q \leq \left(\frac{q-1}{p-1}\right)^{n/2} \|F\|_p. \tag{5.3}$$

Moreover, for any $n \geq 1$ and $1 < p < \infty$, the orthogonal projection onto the nth Wiener chaos J_n is bounded in $L^p(\Omega)$, and

$$\|J_n F\|_p \leq \begin{cases} (p-1)^{n/2}\|F\|_p & \text{if } p > 2, \\ (p-1)^{-n/2}\|F\|_p & \text{if } p < 2. \end{cases} \tag{5.4}$$

In fact, suppose first that $p > 2$ and let $t > 0$ be such that $p - 1 = e^{2t}$. Using the hypercontractivity property with exponents p and 2, we obtain

$$\|J_n F\|_p = e^{nt}\|T_t J_n F\|_p \leq e^{nt}\|J_n F\|_2 \leq e^{nt}\|F\|_2 \leq e^{nt}\|F\|_p.$$

If $p < 2$, we have

$$\|J_n F\|_p = \sup_{\|G\|_q \leq 1} E((J_n F)G) \leq \|F\|_p \sup_{\|G\|_q \leq 1} \|J_n G\|_q \leq e^{nt}\|F\|_p,$$

where q is the conjugate of p, and $q - 1 = e^{2t}$.

As an application we can establish the following lemma.

Lemma 5.1.4 *Fix an integer $k \geq 1$ and a real number $p > 1$. Then, there exists a constant $c_{p,k}$ such that, for any random variable $F \in \mathbb{D}^{k,2}$,*

$$\|E(D^k F)\|_{H^{\otimes k}} \leq c_{p,k}\|F\|_p.$$

Proof Suppose that $F = \sum_{n=0}^{\infty} I_n(f_n)$. Then, by Stroock's formula (4.6), $E(D^k F) = k! f_k$. Therefore,

$$\|E(D^k F)\|_{H^{\otimes k}} = k!\|f_k\|_{H^{\otimes k}} = \sqrt{k!}\|J_k F\|_2.$$

From (5.3) we obtain

$$\|J_k F\|_2 \leq ((p-1) \wedge 1)^{-k/2} \|J_k F\|_p.$$

Finally, applying (5.4) we get

$$\|J_k F\|_p \leq (p-1)^{\text{sign}(p-2)k/2} \|F\|_p,$$

which concludes the proof.   □

The next result can be regarded as a regularizing property of the Ornstein–Uhlenbeck semigroup.

Proposition 5.1.5 *Let $F \in L^p(\Omega)$ for some $p > 1$. Then, for any $t > 0$, we have that $T_t F \in \mathbb{D}^{1,p}$ and there exists a constant c_p such that*

$$\|DT_t F\|_{L^p(\Omega;H)} \leq c_p t^{-1/2} \|F\|_p. \tag{5.5}$$

Proof Consider a sequence of smooth and cylindrical random variables $F_n \in S$ which converges to F in $L^p(\Omega)$. We know that $T_t F_n$ converges to $T_t F$ in $L^p(\Omega)$. We have $T_t F_n \in \mathbb{D}^{1,p}$, and using Mehler's formula (5.1), we can write

$$D(T_t(F_n - F_m)) = D\left(\mathrm{E}'((F_n - F_m)(e^{-t}B + \sqrt{1 - e^{-2t}})B')\right)$$

$$= \frac{e^{-t}}{\sqrt{1 - e^{-2t}}} \, \mathrm{E}'\left(D'((F_n - F_m)(e^{-t}B + \sqrt{1 - e^{-2t}})B'))\right).$$

Then, Lemma 5.1.4 implies that

$$\|D(T_t(F_n - F_m))\|_{L^p(\Omega;H)} \leq c_{p,1} \frac{e^{-t}}{\sqrt{1 - e^{-2t}}} \|F_n - F_m\|_p.$$

Hence, $DT_t F_n$ is a Cauchy sequence in $L^p(\Omega; H)$. Therefore, $T_t F \in \mathbb{D}^{1,p}$ and $DT_t F$ is the limit in $L^p(\Omega; H)$ of $DT_t F_n$. The estimate (5.5) follows by the same arguments.   □

With the above ingredients, we can show an extension of Corollary 4.2.5 to any $p > 1$.

Proposition 5.1.6 *Let $F_n \in \mathbb{D}^{1,p}$ be a sequence of random variables converging to F in $L^p(\Omega)$ for some $p > 1$. Suppose that*

$$\sup_n \|F_n\|_{1,p} < \infty.$$

Then $F \in \mathbb{D}^{1,p}$.

Proof The assumptions imply that there exists a subsequence $(F_{n(k)})_{k\geq 1}$ such that the sequence of derivatives $(DF_{n(k)})_{k\geq 1}$ converges in the weak topology of $L^q(\Omega; H)$ to some element $\alpha \in L^q(\Omega; H)$, where $1/p + 1/q = 1$. By Proposition 5.1.5, for any $t > 0$, we have that $T_t F$ belongs to $\mathbb{D}^{1,p}$ and $DT_t F_{n(k)}$ converges to $DT_t F$ in $L^p(\Omega; H)$. Then, for any $\beta \in L^q(\Omega; H)$, we can write

$$\mathrm{E}(\langle DT_t F, \beta \rangle_H) = \lim_{k\to\infty} \mathrm{E}(\langle DT_t F_{n(k)}, \beta \rangle_H) = \lim_{k\to\infty} e^{-t} \mathrm{E}(\langle T_t DF_{n(k)}, \beta \rangle_H)$$

$$= \lim_{k\to\infty} e^{-t} \mathrm{E}(\langle DF_{n(k)}, T_t \beta \rangle_H) = e^{-t} \mathrm{E}(\langle \alpha, T_t \beta \rangle_H)$$

$$= \mathrm{E}(\langle e^{-t} T_t \alpha, \beta \rangle_H).$$

Therefore, $DT_t F = e^{-t} T_t \alpha$. This implies that $DT_t F$ converges to α as $t \downarrow 0$ in $L^p(\Omega; H)$. Using that D is a closed operator, we conclude that $F \in \mathbb{D}^{1,p}$ and $DF = \alpha$. $\qquad\qquad\square$

5.2 Generator of the Ornstein–Uhlenbeck Semigroup

The generator of the Ornstein–Uhlenbeck semigroup in $L^2(\Omega)$ is the operator given by

$$LF = \lim_{t\downarrow 0} \frac{T_t F - F}{t},$$

and the domain of L is the set of random variables $F \in L^2(\Omega)$ for which the above limit exists in $L^2(\Omega)$. It is easy to show (see Exercise 5.2) that a random variable $F = \sum_{n=0}^{\infty} I_n(f_n)$, $f_n \in L_s^2(\mathbb{R}_+^n)$, belongs to the domain of L if and only if

$$\sum_{n=1}^{\infty} n^2 \|I_n(f_n)\|_2^2 < \infty;$$

and, in this case, $LF = \sum_{n=1}^{\infty} -n I_n(f_n)$. Thus, $\mathrm{Dom}\, L$ coincides with the space $\mathbb{D}^{2,2}$.

We also define the operator L^{-1}, which is the pseudo-inverse of L, as follows. For every $F \in L^2(\Omega)$, set

$$LF = -\sum_{n=1}^{\infty} \frac{1}{n} I_n(f_n).$$

Note that L^{-1} is an operator with values in $\mathbb{D}^{2,2}$ and that $LL^{-1}F = F - \mathrm{E}(F)$, for any $F \in L^2(\Omega)$, so L^{-1} acts as the inverse of L for centered random variables.

Consider now the operator $C = -\sqrt{-L}$ defined by

$$CF = -\sum_{n=1}^{\infty} \sqrt{n} I_n(f_n).$$

The operator C is the infinitesimal generator of the Cauchy semigroup of operators given by

$$Q_t F = \sum_{n=0}^{\infty} e^{-\sqrt{n}t} I_n(f_n).$$

Observe that Dom $C = \mathbb{D}^{1,2}$, as, for any $F \in$ Dom C, we have

$$E(|CF|^2) = \sum_{n=1}^{\infty} n\|I_n(f_n)\|_2^2 = E(\|DF\|_H^2).$$

The next proposition explains the relationship between the operators D, δ, and L.

Proposition 5.2.1 *Let $F \in L^2(\Omega)$. Then, $F \in$ Dom L if and only if $F \in \mathbb{D}^{1,2}$ and $DF \in$ Dom δ and, in this case, we have*

$$\delta DF = -LF.$$

Proof Let $F = \sum_{n=0}^{\infty} I_n(f_n)$. Suppose first that $F \in \mathbb{D}^{1,2}$ and $DF \in$ Dom δ. Then, for any random variable $G = I_m(g_m)$, we have, using the duality relationship (Proposition 3.2.3),

$$E(G\delta DF) = E(\langle DG, DF \rangle_H) = mm! \langle g_m, f_m \rangle_{L^2(\mathbb{R}_+^m)} = E(GmI_m(f_m)).$$

Therefore, the projection of δDF onto the mth Wiener chaos is equal to $mI_m(f_m)$. This implies that the series $\sum_{n=1}^{\infty} nI_n(f_n)$ converges in $L^2(\Omega)$ and its sum is δDF. Therefore, $F \in$ Dom L and $LF = -\delta DF$.

Conversely, suppose that $F \in$ Dom L. Clearly, $F \in \mathbb{D}^{1,2}$. Then, for any random variable $G \in \mathbb{D}^{1,2}$ with Wiener chaos expansion $G = \sum_{n=0}^{\infty} I_n(g_n)$, we have

$$E(\langle DG, DF \rangle_H) = \sum_{n=1}^{\infty} nn! \langle g_n, f_n \rangle_{L^2(\mathbb{R}_+^n)} = -E(GLF).$$

As a consequence, DF belongs to the domain of δ and $\delta DF = -LF$. $\qquad\square$

The operator L behaves as a second-order differential operator on smooth random variables.

Proposition 5.2.2 *Suppose that $F = (F^1, \ldots, F^m)$ is a random vector whose components belong to $\mathbb{D}^{2,4}$. Let φ be a function in $C^2(\mathbb{R}^m)$ with bounded first and second partial derivatives. Then, $\varphi(F) \in \text{Dom } L$ and*

$$L(\varphi(F)) = \sum_{i,j=1}^{m} \frac{\partial^2 \varphi}{\partial x_i \partial x_j}(F)\langle DF^i, DF^j \rangle_H + \sum_{i=1}^{m} \frac{\partial \varphi}{\partial x_i}(F)LF^i.$$

Proof By the chain rule (see Proposition 3.3.2), $\varphi(F)$ belongs to $\mathbb{D}^{1,2}$ and

$$D(\varphi(F)) = \sum_{i=1}^{m} \frac{\partial \varphi}{\partial x_i}(F)DF^i.$$

Moreover, by Proposition 5.2.1, $\varphi(F)$ belongs to $\text{Dom } L$ and $L(\varphi(F)) = -\delta(D(\varphi(F)))$. Using Exercise 3.7 yields the result.   □

In the finite-dimensional case ($\Omega = \mathbb{R}^n$ equipped with the standard Gaussian law), $L = \Delta - x \cdot \nabla$ coincides with the generator of the Ornstein–Uhlenbeck process $(X_t)_{t\geq 0}$ in $\mathbb{R}^n$, which is the solution to the stochastic differential equation

$$dX_t = \sqrt{2}dB_t - X_t dt,$$

where $(B_t)_{t\geq 0}$ is an n-dimensional Brownian motion.

5.3 Meyer's Inequality

The next theorem provides an estimate for the $L^p(\Omega)$-norm of the divergence operator for any $p > 1$. Its proof if due to Pisier (1988) and is based on the boundedness in $L^p(\Omega)$ of the Riesz transform.

Theorem 5.3.1 *For any $p > 1$, there exists a constant $c_p > 0$ such that for any $u \in \mathbb{D}^{1,p}(H)$,*

$$E(|\delta(u)|^p) \leq c_p \Big(E(\|Du\|^p_{L^2(\mathbb{R}^2_+)}) + \|E(u)\|^p_H \Big). \tag{5.6}$$

Proof We can assume that $E(u) = 0$. Indeed, if $\tilde{u} = u - E(u)$ then

$$\|\delta(u)\|_p \leq \|\delta(\tilde{u})\|_p + \|\delta(E(u))\|_p,$$

and, $\delta(E(u)) = I_1(E(u))$ being a Gaussian random variable, we have

$$\|\delta(E(u))\|_p = c_p\|\delta(E(u))\|_2 = c_p\|E(u)\|_H.$$

Let $G \in \mathcal{S}$ be a smooth and cylindrical random variable such that $\|G\|_q \leq 1$, where $1/p + 1/q = 1$. Then

$$E(\delta(u)G) = E(\langle u, DG \rangle_H) = - \int_0^\infty E(\langle LT_t u, DG \rangle_H) dt$$

$$= \int_0^\infty E(\langle \delta DT_t u, DG \rangle_H) dt = \int_0^\infty E\left(\langle DT_t u, D^2 G \rangle_{H^{\otimes 2}}\right) dt$$

$$= \int_0^\infty e^{-t} E\left(\langle T_t Du, D^2 G \rangle_{H^{\otimes 2}}\right) dt$$

$$= \int_0^\infty e^{-t} E\left(\langle Du, T_t D^2 G \rangle_{H^{\otimes 2}}\right) dt$$

$$\leq \|Du\|_{L^p(\Omega;H^{\otimes 2})} \left\| \int_0^\infty e^{-t} T_t D^2 G dt \right\|_{L^q(\Omega;H^{\otimes 2})}.$$

By Mehler's formula (5.1),

$$T_t D^2 G = \tilde{E}\left((D^2 G)(e^{-t} B + \sqrt{1 - e^{-2t}} \tilde{B})\right)$$

$$= \frac{1}{1 - e^{-2t}} \tilde{E}\left(\tilde{D}^2(G(e^{-t} B + \sqrt{1 - e^{-2t}} \tilde{B}))\right).$$

Therefore

$$\int_0^\infty e^{-t} T_t D^2 G dt = \int_0^\infty \frac{e^{-t}}{1 - e^{-2t}} \tilde{E}\left(\tilde{D}^2(G(e^{-t} B + \sqrt{1 - e^{-2t}} \tilde{B}))\right) dt$$

$$= \int_0^1 \frac{1}{1 - x^2} \tilde{E}\left(\tilde{D}^2(G(xB + \sqrt{1 - x^2} \tilde{B}))\right) dx$$

$$= \int_0^{\pi/2} \frac{1}{\sin \theta} \tilde{E}\left(\tilde{D}^2(G(R_\theta(B, \tilde{B})))\right) d\theta,$$

where $R_\theta(B, \tilde{B}) = (\cos \theta)B + (\sin \theta)\tilde{B}$. Since G is a smooth and cylindrical random variable, we can write

$$\int_0^\infty e^{-t} T_t D^2 G dt = \frac{1}{2} \tilde{E}\left(\tilde{D}^2\left(\int_{-\pi/2}^{\pi/2} \frac{1}{\sin \theta} \left(G(R_\theta(B, \tilde{B})) - G(R_{-\theta}(B, \tilde{B}))\right) d\theta\right)\right).$$

From Lemma 5.1.4, we can write

$$\left\| \int_0^\infty e^{-t} T_t D^2 G dt \right\|_{H^{\otimes 2}}^q$$

$$\leq 2^{-q} c_{q,2}^q \tilde{E}\left(\left| \int_{-\pi/2}^{\pi/2} \frac{1}{\sin \theta} \left(G(R_\theta(B, \tilde{B})) - G(R_{-\theta}(B, \tilde{B}))\right) d\theta \right|^q\right),$$

which implies that

$$E\left(\left\|\int_0^\infty e^{-t}T_t D^2 G dt\right\|_{H^{\otimes 2}}^q\right)$$
$$\leq 2^{-q}c_{q,2}^q\, E\tilde{E}\left(\left|\int_{-\pi/2}^{\pi/2}\frac{1}{\sin\theta}\left(G(R_\theta(B,\tilde{B})) - G(R_{-\theta}(B,\tilde{B}))\right)d\theta\right|^q\right).$$

Since the function $1/\sin\theta - 1/\theta$ is integrable in $[0, \pi/2]$, we have

$$E\tilde{E}\left(\left|\int_{-\pi/2}^{\pi/2}\left(\frac{1}{\sin\theta} - \frac{1}{\theta}\right)\left(G(R_\theta(B,\tilde{B})) - G(R_{-\theta}(B,\tilde{B}))\right)d\theta\right|^q\right) \leq c_q E(|G|^q).$$

Therefore, it suffices to study the term

$$E\tilde{E}\left(\left|\int_{-\pi/2}^{\pi/2}\frac{1}{\theta}\left(G(R_\theta(B,\tilde{B})) - G(R_{-\theta}(B,\tilde{B}))\right)d\theta\right|^q\right).$$

For any $\xi \in \mathbb{R}$ we define the transformation

$$T_\xi(B,\tilde{B}) = \begin{pmatrix} \cos\xi & \sin\xi \\ -\sin\xi & \cos\xi \end{pmatrix}\begin{pmatrix} B \\ \tilde{B} \end{pmatrix},$$

which preserves the law of $(B, \tilde{B})$. Moreover, $R_\theta(T_\xi(B,\tilde{B})) = R_{\theta+\xi}(B,\tilde{B})$. Therefore,

$$E\tilde{E}\left(\left|\int_{-\pi/2}^{\pi/2}\frac{1}{\theta}\left(G(R_\theta(B,\tilde{B})) - G(R_{-\theta}(B,\tilde{B}))\right)d\theta\right|^q\right)$$
$$= \frac{1}{\pi}E\tilde{E}\left(\int_{-\pi/2}^{\pi/2}\left|\int_{-\pi/2}^{\pi/2}\frac{1}{\theta}\left(G(R_{\xi+\theta}(B,\tilde{B})) - G(R_{\xi-\theta}(B,\tilde{B})\right)d\theta\right|^q d\xi\right),$$

which is bounded by a constant times $E(|G|^q)$, owing to the boundedness in $L^q(\mathbb{R})$ of the Hilbert transform. $\qquad\square$

As a consequence of Theorem 5.3.1, the divergence operator is continuous from $\mathbb{D}^{1,p}(H)$ to $L^p(\Omega)$, and so we have *Meyer's inequality*:

$$E(|\delta(u)|^p) \leq c_p\left(E(\|Du\|_{L^2(\mathbb{R}_+^2)}^p) + E(\|u\|_H^p)\right) = c_p\|u\|_{1,p,H}^p. \qquad (5.7)$$

Meyer's inequality was first proved in Meyer (1984). It can be extended as follows (see Nualart, 2006, Proposition 1.5.7).

Theorem 5.3.2 *For any $p > 1$, $k \geq 1$, and $u \in \mathbb{D}^{k,p}(H)$,*

$$\|\delta(u)\|_{k-1,p} \leq c_{k,p}\left(E(\|D^k u\|_{L^2(\mathbb{R}_+^{k+1})}^p) + E(\|u\|_H^p)\right) = c_{k,p}\|u\|_{k,p,H}^p.$$

This implies that the operator δ is continuous from $\mathbb{D}^{k,p}(H)$ into $\mathbb{D}^{k-1,p}(H)$.

5.4 Integration-by-Parts Formula

The following integration-by-parts formula will play an important role in the applications of Malliavin calculus to normal approximations.

Proposition 5.4.1 *Let $F \in \mathbb{D}^{1,2}$ with $\mathrm{E}(F) = 0$. Let f be a continuously differentiable function with bounded derivative. Then*

$$\mathrm{E}(f(F)F) = \mathrm{E}\left(f'(F)\langle DF, -DL^{-1}F \rangle_H\right). \tag{5.8}$$

Proof By Proposition 5.2.1,

$$F = LL^{-1}F = -\delta(DL^{-1}F).$$

Then, using the duality relationship (Proposition 3.2.3) and the chain rule (Proposition 3.3.2), we can write

$$\begin{aligned}
\mathrm{E}(f(F)F) &= -\mathrm{E}\left(f(F)\delta(DL^{-1}F)\right) \\
&= -\mathrm{E}\left(\langle D(f(F)), DL^{-1}F \rangle_H\right) \\
&= \mathrm{E}\left(f'(F)\langle DF, -DL^{-1}F \rangle_H\right).
\end{aligned}$$

This completes the proof. $\qquad\qquad\square$

If F belongs to the Wiener chaos $\mathcal{H}_q$, with $q \geq 1$, then $DL^{-1}F = -(1/q)DF$ and

$$\mathrm{E}(f(F)F) = \frac{1}{q}\mathrm{E}\left(f'(F)\|DF\|_H^2\right).$$

Define, for $(\mathrm{P} \circ F^{-1})$-almost all x, the function g_F as follows:

$$g_F(x) = \mathrm{E}\left(\langle DF, -DL^{-1}F \rangle_H \big| F = x\right). \tag{5.9}$$

Then, for any $f \in C_b^1(\mathbb{R})$, we have

$$\mathrm{E}(f(F)F) = \mathrm{E}(f'(F)g_F(F)). \tag{5.10}$$

Moreover, $g_F(F) \geq 0$ almost surely. Indeed, taking $f(x) = \int_0^x \varphi(y)dy$, where φ is smooth and nonnegative, we obtain

$$\mathrm{E}\left(\mathrm{E}\left(\langle DF, -DL^{-1}F \rangle_H \big| F\right)\varphi(F)\right) \geq 0,$$

because $xf(x) \geq 0$.

Ornstein–Uhlenbeck Semigroup

5.5 Nourdin–Viens Density Formula

As a consequence of the integration-by-parts formula (5.10), we have the following formula for the density of a centered random variable, obtained by Nourdin and Viens (2009).

If p is a probability density on $\mathbb{R}$, we define the support of p as the closure of the set $\{x \in \mathbb{R} : p(x) > 0\}$.

Theorem 5.5.1 *Let $F \in \mathbb{D}^{1,2}$ with $\mathrm{E}(F) = 0$. Then, the law of F has density p if and only if $g_F(F) > 0$ a.s., where $g_F(F)$ is the function defined in (5.9). In this case, the support of p is a closed interval containing zero and, for almost all x in the support of p,*

$$p(x) = \frac{\mathrm{E}(|F|)}{2g_F(x)} \exp\left(-\int_0^x \frac{y}{g_F(y)} dy\right).$$ (5.11)

Proof We will show only that if F has density p and support $\mathbb{R}$ then $g_F(F) > 0$ a.s., and formula (5.11) holds. For the general case and the converse statement see Nourdin and Viens (2009). Let $\phi \in C_0^\infty(\mathbb{R})$ and $\Phi' = \phi$. Then, by (5.10),

$$\mathrm{E}(\phi(F)g_F(F)) = \mathrm{E}(\Phi(F)F) = \int_{\mathbb{R}} \Phi(x)xp(x)dx$$

$$= \int_{\mathbb{R}} \left(\int_{-\infty}^x \phi(y)dy\right)xp(x)dx$$

$$= \int_{\mathbb{R}} \phi(y)\left(\int_y^\infty xp(x)dx\right)dy = \int_{\mathbb{R}} \phi(y)\varphi(y)dy,$$

where $\varphi(y) = \int_y^\infty xp(x)dx$. This implies that a.e.

$$\varphi(x) = p(x)g_F(x)$$

and $g_F(x) > 0$ a.e. Taking into account that $\varphi'(x) = -xp(x)$, we obtain

$$\frac{\varphi'(x)}{\varphi(x)} = -\frac{x}{g_F(x)}.$$

Using that $\varphi(0) = \frac{1}{2}\mathrm{E}(|F|)$, and solving the above differential equation, we get

$$\varphi(x) = \frac{1}{2}\mathrm{E}(|F|) \exp\left(-\int_0^x \frac{y}{g_F(y)} dy\right),$$

which completes the proof. $\square$

Corollary 5.5.2 *If there exist $\sigma_{\min}^2, \sigma_{\max}^2 > 0$ such that*

$$\sigma_{\min}^2 \le g_F(F) \le \sigma_{\max}^2$$ (5.12)

a.s., then F has density p. Moreover, the support of p is all $\mathbb{R}$ and, for almost all x in $\mathbb{R}$,

$$\frac{E(|F|)}{2\sigma_{max}^2} \exp\left(-\frac{x^2}{2\sigma_{min}^2}\right) \le p(x) \le \frac{E(|F|)}{2\sigma_{min}^2} \exp\left(-\frac{x^2}{2\sigma_{max}^2}\right).$$

Proof The fact that the support of p is all $\mathbb{R}$ was proved by Nourdin and Viens (2009, Corollary 3.3). The bounds for p follow from (5.11).   □

The random variable $g_F(F)$ can be computed using Mehler's formula (Proposition 5.1.2) and the fact that $-DL^{-1}F = \int_0^\infty e^{-t}T_t(DF)dt$ (see Exercise 5.5). In this way, we obtain

$$g_F(F) = \int_0^\infty e^{-t} E\left(\left\langle DF, E'\left(DF(e^{-t}B + \sqrt{1-e^{-2t}}B')\right)\right\rangle_H \Big| F\right) dt. \quad (5.13)$$

Then Corollary 5.5.2 can be applied if we have uniform upper and lower bounds for $\langle DF, E'(DF(e^{-t}B + \sqrt{1-e^{-2t}}B'))\rangle_H$, as is illustrated in the next example.

Example 5.5.3 Consider the centered random variable

$$F = \max_{t \in [a,b]} X_t - E\left(\max_{t \in [a,b]} X_t\right),$$

where $0 < a < b < \infty$ and $(X_t)_{t \ge 0}$ is a continuous centered Gaussian process (see Example 3.5.2).

It was shown by Kim and Pollard (1990, Lemma 2.6) that if $E(X_t - X_s)^2 \ne 0$ for all $s \ne t$ then X_t attains its maximum in $[a,b]$ almost surely at a unique random point τ (see Lemma 1.8.4 for the Brownian motion case). Moreover, by Exercise 3.4, $F \in \mathbb{D}^{1,2}$ and $D_rF = 1_{[0,\tau]}(r)$.

Therefore, applying formula (5.13), we get

$$g_F(F) = \int_0^\infty e^{-t} E(X_\tau X_{\tau'}) dt,$$

where τ' is the random point where the process $e^{-t}X_t + \sqrt{1-e^{-2t}}X_t'$ attains its maximum in $[a,b]$. Thus, if there exist $\sigma_{min}^2, \sigma_{max}^2 > 0$ such that

$$\sigma_{min}^2 \le E(X_s X_t) \le \sigma_{max}^2, \quad (5.14)$$

for all $s,t \in [a,b]$, then condition (5.12) of Corollary 5.5.2 holds.

For example, if X_t is a fractional Brownian motion B_t^H with Hurst parameter $H \ge 1/2$ (see Example 3.5.3) then (5.14) holds with $\sigma_{min} = a^H$ and $\sigma_{max} = b^H$. In fact,

$$E(B_s^H B_t^H) \le \sqrt{E((B_s^H)^2)} \sqrt{E((B_t^H)^2)} = (st)^H \le b^{2H}$$

and

$$E(B_s^H B_t^H) = H(2H - 1) \int_0^s \int_0^t |v - u|^{2H-2} du dv$$

$$\geq H(2H - 1) \int_0^a \int_0^a |v - u|^{2H-2} du dv = E((B_a^H)^2) = a^{2H}.$$

Exercises

5.1   Show that $(T_t)_{t \geq 0}$ is the semigroup of transition probabilities of a Markov process, with values in $C(\mathbb{R}_+)$, whose invariant measure is the Wiener measure. This process can be expressed in terms of a Wiener sheet W as follows:

$$X_{t,\tau} = \sqrt{2} \int_{-\infty}^t \int_0^\tau e^{-(t-s)} W(d\sigma, ds), \qquad \tau, t \geq 0.$$

5.2   Show that a random variable $F = \sum_{n=0}^\infty I_n(f_n)$, $f_n \in L_s^2(\mathbb{R}_+^n)$, belongs to the domain of L if and only if

$$\sum_{n=1}^\infty n^2 \|I_n(f_n)\|_2^2 < \infty,$$

and, in this case, $LF = \sum_{n=1}^\infty -n I_n(f_n)$.

5.3   Let $F = \exp\left(B(h) - \frac{1}{2} \int_{\mathbb{R}_+} h^2(s) ds\right)$, $h \in H$. Show that

$$LF = -\left(B(h) - \int_{\mathbb{R}_+} h^2(s) ds\right) F.$$

5.4   Show that the operator $(I - L)^{-\alpha}$ is a contraction in $L^p(\Omega)$, for any $p \geq 1$, where $\alpha > 0$.
      Hint: Use the equation $(1 + n)^{-\alpha} = (\Gamma(\alpha))^{-1} \int_0^\infty e^{-(n+1)t} t^{\alpha-1} dt$.

5.5   Let $F \in L^2(\Omega)$ with $E(F) = 0$. Show that

$$L^{-1} F = -\int_0^\infty T_t(F) dt.$$

5.6   Let $(X_t)_{t \in [0,T]}$ be a continuous centered Gaussian process such that there exists constants $\sigma_{min}^2, \sigma_{max}^2 > 0$ satisfying

$$\sigma_{min}^2 \leq E(X_s X_t) \leq \sigma_{max}^2,$$

for all $s, t \in [0, T]$. Let $f \in C^1(\mathbb{R})$ be such that there exist constants $\alpha, \beta > 0$ satisfying $\alpha \leq f'(x) \leq \beta$ for all $x \in \mathbb{R}$. Show that the random variable $F = \int_0^T f(X_s) ds - E\left(\int_0^T f(X_s) ds\right)$ has density p satisfying, a.e. in $\mathbb{R}$,

$$\frac{E(|F|)}{2\beta^2 \sigma_{max}^2 T^2} \exp\left(-\frac{x^2}{2\alpha^2 \sigma_{min}^2 T^2}\right) \leq p(x) \leq \frac{E(|F|)}{2\alpha^2 \sigma_{min}^2 T^2} \exp\left(-\frac{x^2}{2\beta^2 \sigma_{max}^2 T^2}\right).$$

6

Stochastic Integral Representations

This chapter deals with the following problem. Given a random variable F in $L^2(\Omega)$, with $E(F) = 0$, find a stochastic process u in $\text{Dom}\,\delta$ such that $F = \delta(u)$. We present two different answers to this question, both integral representations. The first is the Clark–Ocone formula, given in Ocone (1984), in which u is required to be adapted. Therefore, the process u is unique and its expression involves a conditional expectation of the Malliavin derivative of F. The second uses the inverse of the Ornstein–Uhlenbeck generator. We then present some applications of these integral representations.

6.1 Clark–Ocone formula

Let $B = (B_t)_{t \geq 0}$ be a Brownian motion on a probability space $(\Omega, \mathcal{F}, P)$ such that $\mathcal{F}$ is generated by B, equipped with its Brownian filtration $(\mathcal{F}_t)_{t \geq 0}$. The next result expresses the integrand of the integral representation theorem of a square integrable random variable (Theorem 2.9.1) in terms of the conditional expectation of its Malliavin derivative.

Theorem 6.1.1 (Clark–Ocone formula) *Let $F \in \mathbb{D}^{1,2} \cap L^2(\Omega, \mathcal{F}_T, P)$. Then F admits the following representation:*

$$F = E(F) + \int_0^T E(D_t F | \mathcal{F}_t) dB_t.$$

Proof By Theorem 2.9.1, there exists a unique process $u \in L_T^2(\mathcal{P})$ such that $F \in L^2(\Omega, \mathcal{F}_T, P)$ admits the stochastic integral representation

$$F = E(F) + \int_0^T u_t dB_t.$$

It suffices to show that $u_t = E(D_t F | \mathcal{F}_t)$ for almost all $(t, \omega) \in [0, T] \times \Omega$. Consider a process $v \in L_T^2(\mathcal{P})$. On the one hand, the isometry property yields

$$E(\delta(v)F) = \int_0^T E(v_s u_s)ds.$$

On the other hand, by the duality relationship (Proposition 3.2.3), and taking into account that v is progressively measurable,

$$E(\delta(v)F) = E\left(\int_0^T v_t D_t F dt\right) = \int_0^T E(v_s E(D_t F|\mathcal{F}_t))dt.$$

Therefore, $u_t = E(D_t F|\mathcal{F}_t)$ for almost all $(t, \omega) \in [0, T] \times \Omega$, which concludes the proof.   □

Consider the following simple examples of the application of this formula.

Example 6.1.2   Suppose that $F = B_t^3$. Then $D_s F = 3B_t^2 \mathbf{1}_{[0,t]}(s)$ and

$$E(D_s F|\mathcal{F}_s) = 3E((B_t - B_s + B_s)^2|\mathcal{F}_s) = 3(t - s + B_s^2).$$

Therefore

$$B_t^3 = 3\int_0^t (t - s + B_s^2)dB_s. \tag{6.1}$$

This formula should be compared with Itô's formula,

$$B_t^3 = 3\int_0^t B_s^2 dB_s + 3\int_0^t B_s ds. \tag{6.2}$$

Notice that equation (6.1) contains only a stochastic integral but it is not a martingale, because the integrand depends on t, whereas (6.2) contains two terms and one is a martingale. Moreover, the integrand in (6.1) is unique.

Example 6.1.3   Let $F = \sup_{t\in[0,1]} B_t$. By Exercise 3.4, $F \in \mathbb{D}^{1,2}$ and $D_t F = \mathbf{1}_{[0,\tau]}(t)$, where τ is the unique random point in $[0, 1]$ where B attains its maximum almost surely (see Lemma 1.8.4). By the reflection principle (Theorem 1.8.2),

$$E(\mathbf{1}_{[0,\tau]}(t)|\mathcal{F}_t) = E\left(\mathbf{1}_{\{\sup_{t \le s \le 1}(B_s - B_t) \ge \sup_{0 \le s \le t}(B_s - B_t)\}}\Big|\mathcal{F}_t\right)$$

$$= 2 - 2\Phi\left(\frac{\sup_{0 \le s \le t}(B_s - B_t)}{\sqrt{1 - t}}\right),$$

where Φ denotes the cumulative distribution of the standard normal. Therefore

$$F = E(F) + \int_0^1 \left(2 - 2\Phi\left(\frac{\sup_{0 \le s \le t}(B_s - B_t)}{\sqrt{1 - t}}\right)\right)dB_t.$$

Example 6.1.4 Consider the Brownian motion local time $(L_t^x)_{t \geq 0, x \in \mathbb{R}}$ defined in Section 2.5. For any $\varepsilon > 0$, we set

$$p_\varepsilon(x) = (2\pi\varepsilon)^{-1/2} e^{-x^2/(2\varepsilon)}.$$

We have that, as $\epsilon \to 0$,

$$F_\varepsilon = \int_0^t p_\varepsilon(B_s - x) ds \xrightarrow{L^2(\Omega)} L_t^x. \tag{6.3}$$

In fact, by (2.17),

$$F_\varepsilon - L_t^x = \int_{\mathbb{R}} p_\varepsilon(y - x)(L_t^y - L_t^x) dy.$$

The continuity of the local time in (t, x) implies that $F_\varepsilon - L_t^x$ converges almost surely to zero as $\epsilon \to 0$ for all (t, x). Moreover, by Tanaka's formula (Corollary 2.5.2), for any $p > 2$,

$$\sup_{x \in \mathbb{R}} \|L_t^x\|_p < \infty.$$

Therefore the convergence is also in $L^2(\Omega)$, which shows (6.3). Applying the derivative operator yields

$$D_r F_\epsilon = \int_0^t p_\epsilon'(B_s - x) D_r B_s ds = \int_r^t p_\epsilon'(B_s - x) ds.$$

Thus

$$\mathrm{E}(D_r F_\epsilon | \mathcal{F}_r) = \int_r^t \mathrm{E}(p_\epsilon'(B_s - B_r + B_r - x) | \mathcal{F}_r) ds$$

$$= \int_r^t p_{\epsilon+s-r}'(B_r - x) ds.$$

As a consequence, taking the limit as $\epsilon \to 0$, we obtain the following integral representation of the Brownian local time:

$$L_t^x = \mathrm{E}(L_t^x) + \int_0^t \varphi(t - r, B_r - x) dB_r,$$

where

$$\varphi(r, y) = \int_0^r p_s'(y) ds.$$

In the last part of this section we will discuss how to derive a Clark–Ocone formula after a change of measure. Consider an adapted process $\theta = (\theta_t)_{t \in [0,T]}$ satisfying Novikov's condition (2.24). Define

$$Z_T = \exp\left(-\int_0^T \theta_s dB_s - \frac{1}{2} \int_0^T \theta_s^2 ds\right).$$

Consider the probability Q on $\mathcal{F}_T$ given by the density $dQ/dP = Z_T$. Set

$$\tilde{B}_t = B_t + \int_0^t \theta_s ds, \quad t \in [0, T].$$

By Girsanov's theorem (Theorem 2.10.3), under Q, $\tilde{B} = (\tilde{B}_t)_{t \in [0,T]}$ is a Brownian motion.

In general $\mathcal{F}_T^{\tilde{B}} \subset \mathcal{F}_T^B$, with a strict inclusion. Can we represent an $\mathcal{F}_T^{\tilde{B}}$-measurable random variable as a stochastic integral with respect to the Q-Brownian motion $\tilde{B}$ using the Clark–Ocone formula? The following theorem provides an answer to this question.

Theorem 6.1.5   *Suppose that $F \in \mathbb{D}^{1,2}$, $\theta \in \mathbb{D}^{1,2}(L^2([0, T]))$, and*

- $E(Z_T^2 F^2) + E(Z_T^2 \|DF\|_H^2) < \infty$,
- $E\left(Z_T^2 F^2 \int_0^T \left(\theta_t + \int_t^T D_t \theta_s dB_s + \int_t^T \theta_s D_t \theta_s ds\right)^2 dt\right) < \infty$.

Then

$$F = E_Q(F) + \int_0^T E_Q\left(D_t F + F \int_t^T D_t \theta_s d\tilde{B}_s \Big| \mathcal{F}_t\right) d\tilde{B}_t.$$

We refer to Karatzas and Ocone (1991) for a proof of this result with extension to a multidimensional case and applications to hedging in a generalized Black–Scholes model.

6.2 Modulus of Continuity of the Local Time

As a consequence of the Clark–Ocone formula (Theorem 6.1.1), we have the following central limit theorem for the L^2-modulus of continuity of the local time.

Theorem 6.2.1   *For each fixed $t > 0$, as h tends to zero we have the following convergence in law:*

$$h^{-3/2}\left(\int_{\mathbb{R}} (L_t^{x+h} - L_t^x)^2 dx - 4th\right) \xrightarrow{\mathcal{L}} 8\sqrt{\frac{\alpha_t}{3}}\eta,$$

where

$$\alpha_t = \int_{\mathbb{R}} (L_t^x)^2 dx \tag{6.4}$$

and η is an $N(0, 1)$ random variable that is independent of B.

Theorem 6.2.1 was proved by Chen *et al.* (2010) using the method of moments. We give below the main ideas of a proof, based on the Clark–Ocone formula, due to Hu and Nualart (2009).

We make use of the notation, for $t, h > 0$,

$$G_{t,h} = \int_{\mathbb{R}} (L_t^{x+h} - L_t^x)^2 dx.$$

We first establish some preliminary results on the stochastic integral representation of the random variable $G_{t,h}$, obtained using the Clark–Ocone formula.

The random variable α_t appearing in (6.4) can be expressed in terms of the self-intersection local time of Brownian motion. In fact,

$$\alpha_t = \int_{\mathbb{R}} \left(\int_0^t \delta_x(B_s) ds \right)^2 dx = 2 \int_0^t \int_0^v \delta_0(B_v - B_u) du dv.$$

In the same way,

$$G_{t,h} = -2 \int_0^t \int_0^v \left(\delta(B_v - B_u + h) + \delta(B_v - B_u - h) - 2\delta(B_v - B_u) \right) du dv.$$

Applying the Clark–Ocone formula (Theorem 6.1.1), we can derive the following stochastic integral representation for $G_{t,h}$.

Proposition 6.2.2 *We have*

$$G_{t,h} = \mathrm{E}(G_{t,h}) - 4 \int_0^t u_{t,h}(r) dB_r - 4 \int_0^t \Psi_h(r) dB_r, \qquad (6.5)$$

where

$$u_{t,h}(r) = \int_0^r \int_0^h (p_{t-r}(B_r - B_u + y) - p_{t-r}(B_r - B_u - y)) dy du$$

and

$$\Psi_h(r) = \int_0^r (\mathbf{1}_{[0,h]}(B_r - B_u) - \mathbf{1}_{[0,h]}(B_u - B_r)) du.$$

Proof Let $\epsilon > 0$. For $u < r < v$, we have

$$E\left(D_r(p_\epsilon(B_v - B_u + h) + p_\epsilon(B_v - B_u - h) - 2p_\epsilon(B_v - B_u))\middle|\mathcal{F}_r\right)$$
$$= E\left(p_\epsilon'(B_v - B_u + h) + p_\epsilon'(B_v - B_u - h) - 2p_\epsilon'(B_v - B_u)\middle|\mathcal{F}_r\right)$$
$$= p_{\epsilon+v-r}'(B_r - B_u + h) + p_{\epsilon+v-r}'(B_r - B_u - h) - 2p_{\epsilon+v-r}'(B_r - B_u)$$
$$= \int_0^h \left(p_{\epsilon+v-r}''(B_r - B_u + y) - p_{\epsilon+v-r}''(B_r - B_u - y)\right) dy$$
$$= 2\int_0^h \left(\frac{\partial p_{\epsilon+v-r}}{\partial v}(B_r - B_u + y) - \frac{\partial p_{\epsilon+v-r}}{\partial v}(B_r - B_u - y)\right) dy,$$

where $p_\epsilon = (2\pi\epsilon)^{-1/2}\exp(-x^2/(2\epsilon))$. Integrating in u and v and letting $\epsilon \to 0$ yields the result. $\qquad\square$

Proof of Theorem 6.2.1 The proof will be carried out in several steps.

Step 1 It was proved by Chen *et al.* (2010, Lemma 8.1) that $E(G_{t,h}) = 4th + O(h^2)$. Therefore, it suffices to show that, as h tends to zero,

$$h^{-3/2}(G_{t,h} - E(G_{t,h})) \xrightarrow{\mathcal{L}} 8\sqrt{\frac{\alpha_t}{3}}\eta, \qquad (6.6)$$

where η is an $N(0, 1)$ random variable that is independent of B.

Step 2 The stochastic integral $\int_0^t u_{t,h}(r)dB_r$ appearing in (6.5) makes no contribution to the limit (6.6). That is,

$$h^{-3/2}\int_0^t u_{t,h}(r)dB_r$$

converges in $L^2(\Omega)$ to zero as h tends to zero. This follows from the estimate

$$E\left(\int_0^t |u_{t,h}(r)|^2 dr\right) \le Ch^4, \qquad (6.7)$$

for all $h > 0$.

Proof of (6.7) We can write

$$E\left(|u_{t,h}(r)|^2\right) = \int_0^r \int_0^r \int_0^h \int_0^h \int_{-\eta_1}^{\eta_1} \int_{-\eta_2}^{\eta_2} E(p_{t-r}'(B_r - B_{u_1} + \xi_1)$$
$$\times p_{t-r}'(B_r - B_{u_2} + \xi_2))\, d\xi_1 d\xi_2 d\eta_1 d\eta_2 du_1 du_2.$$

By a symmetry argument, it suffices to integrate over the region $0 < u_1 <$

$u_2 < r$. In this region,

$$E(p'_{t-r}(B_r - B_{u_1} + \xi_1)p'_{t-r}(B_r - B_{u_2} + \xi_2))$$
$$= E\left(p'_{t-r+u_2-u_1}(B_r - B_{u_2} + \xi_1)p'_{t-r}(B_r - B_{u_2} + \xi_2)\right)$$
$$\leq \|p_{r-u_2}\|_{p_1}\|p'_{t-r+u_2-u_1}\|_{p_2}\|p'_{t-r}\|_{p_3},$$

where $1/p_1 + 1/p_2 + 1/p_3 = 1$. It is easy to see that

$$\|p_{r-u_2}\|_{p_1} \leq C(r - u_2)^{-1/2+1/(2p_1)},$$
$$\|p'_{t-r+u_2-u_1}\|_{p_2} \leq C(t - r + u_2 - u_1)^{-1+1/(2p_2)} \leq C(u_2 - u_1)^{-1+1/(2p_2)},$$
$$\|p'_{t-r}\|_{p_3} \leq C(t - r)^{-1+1/(2p_3)},$$

for some constant $C > 0$. Thus

$$E\left(|u_{t,h}(r)|^2\right) \leq C \int_0^r \int_0^{u_2} \int_0^h \int_0^h \int_{-\eta_1}^{\eta_1} \int_{-\eta_2}^{\eta_2} (r - u_2)^{-1/2+1/(2p_1)}$$
$$\times (u_2 - u_1)^{-1+1/(2p_2)}(t - r)^{-1+1/(2p_3)}d\xi_1 d\xi_2 d\eta_1 d\eta_2 du_1 du_2$$
$$\leq Ch^4,$$

which concludes the proof of (6.7).

Step 3   Consider the martingale

$$M_t^h = h^{-3/2} \int_0^t \Psi_h(r)dB_r.$$

In order to show Theorem 6.2.1, it suffices to prove the following convergence in law as h tends to zero:

$$M_t^h \xrightarrow{\mathcal{L}} 2\eta\sqrt{\frac{\alpha_t}{3}},$$

where η is a standard normal random variable that is independent of B.

From the asymptotic version of the Ray–Knight theorem (see Revuz and Yor, 1999, Theorem (2.3), p. 524) it suffices to show that, for any $t \geq 0$,

$$\langle M^h \rangle_t = h^{-3} \int_0^t \Psi_h(r)^2 dr \xrightarrow{L^2(\Omega)} \tfrac{4}{3}\alpha_t \qquad (6.8)$$

and

$$\sup_{0 \leq s \leq t} |\langle M^h, B \rangle_s| = \sup_{0 \leq s \leq t} h^{-3/2}\left|\int_0^s \Psi_h(r)dr\right| \xrightarrow{L^2(\Omega)} 0. \qquad (6.9)$$

In fact, (6.8) and (6.9) imply that (B, β^h) converges in law to (B, β), where β is a Brownian motion that is independent of B, and β^h is such that $M_t^h = \beta^h(\langle M^h \rangle_t)$ (the asymptotic version of the Ray–Knight theorem).

Proof of (6.8) By the occupation formula (2.17), we have

$$\Psi_h(r) = \int_{\mathbb{R}} (\mathbf{1}_{[0,h]}(B_r - x) - \mathbf{1}_{[0,h]}(x - B_r))L_r^x dx$$

$$= \int_0^h \left(L_r^{B_r-y} - L_r^{B_r+y}\right) dy.$$

Tanaka's formula (Theorem 2.5.1) for the Brownian motion $\{B_r - B_s, 0 \le s \le r\}$ yields

$$L_r^{B_r-y} - L_r^{B_r+y} = -2y - 2(B_r - y)^+ + 2(B_r + y)^+$$

$$+ 2\int_0^r \mathbf{1}_{\{y > |B_r - B_s|\}} \hat{d}B_s,$$

where $\hat{d}B_s$ denotes the backward Itô stochastic integral (see Section 2.8). Then

$$\Psi_h(r) = -h^2 + 2\int_0^h \left[(B_r + y)^+ - (B_r - y)^+\right] dy$$

$$+ 2\int_0^r (h - B_r - B_s|)^+ \hat{d}B_s. \tag{6.10}$$

Therefore, it suffices to show that

$$4h^{-3} \int_0^t \left(\int_0^r (h - |B_r - B_s|)^+ \hat{d}B_s\right)^2 dr \xrightarrow{L^2(\Omega)} \tfrac{4}{3}\alpha_t.$$

By Itô's formula (Theorem 2.4.3), we can write

$$\left(\int_0^r (h - |B_r - B_s|)^+ \hat{d}B_s\right)^2$$

$$= 2\int_0^r \left(\int_s^r (h - |B_r - B_u|)^+ \hat{d}B_u\right)(h - |B_r - B_s|)^+ \hat{d}B_s$$

$$+ \int_0^r ((h - |B_r - B_s|)^+)^2 ds$$

$$= I_1(r, h) + I_2(r, h).$$

It is not difficult to show that the term $I_1(r, h)$ does not contribute to the above limit. Therefore, we need only to show that

$$h^{-3} \int_0^t \int_0^r ((h - |B_r - B_s|)^+)^2 ds\,dr \xrightarrow{L^2(\Omega)} \tfrac{1}{3}\alpha_t.$$

This follows from

$$\alpha_t = 2\int_0^t \int_{\mathbb{R}} L_r^x L_{dr}^x dx = 2\int_0^t L_r^{B_r} dr,$$

$$\int_0^t \int_0^r ((h - |B_r - B_s|)^+)^2 ds\,dr = \int_0^t \int_{\mathbb{R}} ((h - |B_r - x|)^+)^2 L_r^x dx\,dr,$$

and

$$\int_{\mathbb{R}} \frac{((h - |x|)^+)^2}{h^3} dx = \tfrac{2}{3}.$$

This concludes the proof of (6.8).

Proof of (6.9) In view of (6.10) it suffices to show that

$$\sup_{0 \le t \le t_1} \left| h^{-3/2} \int_0^t \left(\int_0^r (h - |B_r - B_s|)^+ dB_s \right) dr \right|$$

converges to zero in $L^2(\Omega)$ as h tends to zero, for any $t_1 > 0$. For any $p \ge 2$ and any $0 \le s < t$, we can write, by Fubini's theorem and the Burkholder–Davis–Gundy inequality (Theorem 2.2.3),

$$E\left(\left| \int_s^t \left(\int_0^r (h - |B_r - B_v|)^+ dB_v \right) dr \right|^p \right)$$

$$\le c_p \left\{ E\left(\left| \int_0^s \left(\int_s^t (h - |B_r - B_v|)^+ dr \right)^2 dv \right|^{p/2} \right) \right.$$

$$\left. + E\left(\left| \int_s^t \left(\int_v^t (h - |B_r - B_v|)^+ dr \right)^2 dv \right|^{p/2} \right) \right\}$$

$$= c_p (I_1 + I_2).$$

The term I_1 can be expressed using the occupation formula (2.17) as follows:

$$I_1 = E\left(\left| \int_0^s \left(\int_{\mathbb{R}} (h - |x - B_v|)^+ (L_t^x - L_s^x) dx \right)^2 dv \right|^{p/2} \right)$$

$$\le s^{p/2} h^{2p} E\left(\sup_x |L_t^x - L_s^x|^p \right).$$

By the inequalities for local time proved, for instance, in Barlow and Yor (1981), we obtain

$$I_1 \le c_p h^{2p} |t - s|^{p/2}.$$

Similarly,

$$I_2 = E\left(\left| \int_s^t \left(\int_{\mathbb{R}} (h - |x - B_v|)^+ (L_t^x - L_v^x) dx \right)^2 dv \right|^{p/2} \right)$$

$$\le h^{2p} |t - s|^{p/2} \sup_{s \le v \le t} E\left(\sup_x |L_t^x - L_v^x|^p \right)$$

$$\le c_p h^{2p} |t - s|^p.$$

Finally, a standard application of the Garsia–Rodemich–Rumsey lemma (see Garsia *et al.*, 1970/71) allows us to conclude the proof of (6.9).

The proof of Theorem 6.2.1 is now completed.   □

Theorem 6.2.1 can be extended to the third moment. In fact, the Clark–Ocone formula applied to $h^{-2}\int_{\mathbb{R}}(L_t^{x+h}-L_t^x)^3dx$ gives the following central limit theorem for the third integrated moment of the local time; this was proved by Hu and Nualart (2010).

Theorem 6.2.3   *As h tends to zero,*

$$h^{-2}\int_{\mathbb{R}}(L_t^{x+h}-L_t^x)^3dx \xrightarrow{\mathcal{L}} 8\sqrt{3}\Big(\int_{\mathbb{R}}(L_t^x)^3dx\Big)^{1/2}\eta,$$

where η is an $N(0,1)$ random variable that is independent of B.

6.3 Derivative of the Self-Intersection Local Time

In this section we obtain a stochastic integral representation formula for the derivative of the self-intersection local time of a Brownian motion, defined as the following limit in $L^2(\Omega)$:

$$\gamma_t := \lim_{\epsilon\to 0}\int_0^t\int_0^s p'_\epsilon(B_s-B_u)duds. \tag{6.11}$$

This limit can also be interpreted as the value of the following derivative at zero:

$$\gamma_t = -\frac{d}{dx}\Big(\int_0^t\int_0^s \delta_x(B_s-B_u)duds\Big)\Big|_{x=0}.$$

Proposition 6.3.1   *The limit* (6.11) *exists in $L^2(\Omega)$, and*

$$\gamma_t = 2\int_0^t\Big(\int_0^r p_{t-r}(B_r-B_u)du - L_r^{B_r}\Big)dB_r.$$

Proof   Set $\gamma_t^\epsilon = \int_0^t\int_0^s p'_\epsilon(B_s-B_u)duds$. Then

$$D_r\gamma_t^\epsilon = \int_0^t\int_0^s p''_\epsilon(B_s-B_u)\mathbf{1}_{[u,s]}(r)duds$$

$$= \int_r^t\int_0^r p''_\epsilon(B_s-B_u)duds$$

and

$$E(D_r\gamma_t^\epsilon|\mathcal{F}_r) = \int_r^t \int_0^r p''_{\epsilon+s-r}(B_r - B_u)duds$$

$$= 2\int_r^t \int_0^r \frac{\partial p_{\epsilon+s-r}}{\partial u}(B_r - B_u)duds$$

$$= 2\int_0^r (p_{\epsilon+t-r}(B_r - B_u) - p_\epsilon(B_r - B_u))du.$$

As ϵ tends to zero, this converges in $L^2(\Omega \times [0,t])$ to

$$\int_0^r p_{t-r}(B_r - B_u)du - L_r^{B_r},$$

which implies the result. $\square$

The process γ_t is a Dirichlet process (it has zero quadratic variation and infinite total variation). This process was studied by Rogers and Walsh (1994) and Rosen (2005). The following theorem is due to Rogers and Walsh (1994).

Theorem 6.3.2 *The process γ has a 4/3-variation in $L^2(\Omega)$ given by*

$$\langle\gamma\rangle_{4/3,t} = K\int_0^t \left(L_r^{B_r}\right)^{2/3} dr,$$

where $K = E(|B_1|^{4/3})E\left(\left(\int_\mathbb{R}(L_1^z)^2 dz\right)^{2/3}\right)$.

The proof is based on Gebelein's inequality for Gaussian random variables. An alternative proof by Hu *et al.* (2014a) uses the integral representation and ideas from fractional martingales.

6.4 Application of the Clark–Ocone Formula in Finance

Assume that the price of a risky asset $(S_t)_{t\in[0,T]}$ follows the Black–Scholes model under a risk-neutral probability P:

$$dS_t = rS_t dt + \sigma S_t dB_t, \quad t \in [0,T],$$

where $r > 0$ is the constant interest rate and $\sigma > 0$ is the volatility. By Itô's formula (Theorem 2.4.3), this is equivalent to saying that

$$S_t = S_0 \exp\left(\sigma B_t + \left(r - \frac{\sigma^2}{2}\right)t\right).$$

Fix a time $T > 0$, and consider a payoff $H \geq 0$ which is $\mathcal{F}_T$-measurable and such that $E(H^2) < \infty$. Applying the integral representation theorem (Theorem 2.9.1) to $e^{-rT}H$ yields

$$e^{-rT}H = E(e^{-rT}H) + \int_0^T u_t dB_t,$$

where $u \in L_T^2(\mathcal{P})$. As a consequence, we can show that H can be replicated; that is, there exists a self-financing portfolio with value X_t satisfying $X_T = H$. This implies that this market is complete. A self-financing portfolio is characterized by the value at 0 and by the amount Δ_t of shares in the portfolio at any time $t \in [0, T]$. Its value is given by the equation

$$dX_t = \Delta_t dS_t + r(X_t - \Delta_t S_t)dt.$$

Then, to construct such a portfolio it suffices to take $X_0 = E(e^{-rT}H)$ and $\Delta_t = e^{rt}u_t/(\sigma S_t)$. Indeed, the process $X_t = e^{rt}\left(E(H) + \int_0^t u_s dB_s\right)$ satisfies

$$dX_t = e^{rt}u_t dB_t + re^{rt}X_t dt$$
$$= \Delta_t(dS_t - rS_t dt) + rX_t dt.$$

As a consequence of the Clark–Ocone formula we have the following result.

Proposition 6.4.1 *The hedging portfolio of a derivative security with payoff $H \in L^2(\Omega, \mathcal{F}_T, P)$ is given by*

$$\Delta_t = \frac{e^{-r(T-t)}}{\sigma S_t}E(D_t H|\mathcal{F}_t).$$

Proof The Clark–Ocone formula implies that $u_t = e^{-rT}E(D_t H|\mathcal{F}_t)$ and the result then follows from the previous computations. □

In the case of a "vanilla" option, which by definition has the form $H = \varphi(S_T)$, where φ is a differentiable function such that φ and its derivative φ' have polynomial growth, we have $D_t H = \varphi'(S_T)D_t S_T = \varphi'(S_T)\sigma S_T$ and

$$\Delta_t = \frac{e^{-r(T-t)}}{S_t}E(\varphi'(S_T)S_T|\mathcal{F}_t).$$

By the Markov property (Theorem 1.6.1), this expression can be written in the form $F(t, S_t)$, where

$$F(t, x) = e^{-r(T-t)}E\left(\varphi'\left(x\frac{S_T}{S_t}\right)\frac{S_T}{S_t}\right).$$

Then, one can use the techniques of Malliavin calculus to remove the derivative. This would require one to apply an integration-by-parts formula that will be proved in Chapter 7 (see Proposition 7.2.4). We obtain

$$F(t, x) = \frac{e^{-r(T-t)}}{x\sigma(T-t)} \mathrm{E}\left(\varphi\left(x\frac{S_T}{S_t}\right)(B_T - B_t)\right).$$

This expression is well suited for Monte Carlo simulations (see Fournier *et al.*, 1999, and Kohatsu-Higa and Montero, 2004). By the Markov property, the price of the security is of the form $v(t, S_t)$, where

$$v(t, x) = e^{-r(T-t)} \mathrm{E}\left(\varphi\left(x\frac{S_T}{S_t}\right)\right),$$

and, in that case, $\Delta_t = F(t, S_t) = v_x(t, S_t)$.

6.5 Second Integral Representation

Recall that L is the generator of the Ornstein–Uhlenbeck semigroup introduced in Chapter 5.

Proposition 6.5.1 *Let F be in $\mathbb{D}^{1,2}$ with $\mathrm{E}(F) = 0$. Then the process*

$$u = -DL^{-1}F$$

belongs to Dom δ *and satisfies $F = \delta(u)$. Moreover $u \in L^2(\Omega; H)$ is unique among all square integrable processes with a chaos expansion*

$$u_t = \sum_{q=0}^{\infty} I_q(f_q(t))$$

such that $f_q(t, t_1, \ldots, t_q)$ is symmetric in all $q + 1$ variables $t, t_1, \ldots, t_q$.

Proof By Proposition 5.2.1,

$$F = LL^{-1}F = -\delta(DL^{-1}F).$$

Clearly, the process $u = -DL^{-1}F$ has a Wiener chaos expansion with functions symmetric in all their variables. To show uniqueness, let $v \in L^2(\Omega; H)$ with a chaos expansion $v_t = \sum_{q=0}^{\infty} I_q(g_q(t))$, such that the function $g_q(t, t_1, \ldots, t_q)$ is symmetric in all $q + 1$ variables $t, t_1, \ldots, t_q$ and such that $\delta(v) = F$. Then, there exists a random variable $G \in \mathbb{D}^{1,2}$ such that $DG = v$. Indeed, it suffices to take

$$G = \sum_{q=0}^{\infty} \frac{1}{q+1} I_{q+1}(g_q).$$

We claim that $G = -L^{-1}F$. This follows from $LG = -\delta DG = -\delta(v) = -F$. The proof is now complete. $\square$

It is important to notice that, unlike the Clark–Ocone formula, which requires that the underlying process is a Brownian motion, the representation provided in Proposition 6.5.1 holds in the context of a general Gaussian isonormal process.

6.6 Proving Tightness Using Malliavin Calculus

In this section we summarize an application of the second representation proved in Proposition 6.5.1 to derive tightness in the asymptotic behavior of the self-intersection local time of a fractional Brownian motion. Let $d \geq 2$. The d-dimensional fractional Brownian motion $B^H = (B_t^H)_{t \geq 0}$ with Hurst parameter $H \in (0, 1)$ is a mean-zero Gaussian process with covariance

$$E\left(B_t^{H,i} B_s^{H,j}\right) = \tfrac{1}{2}\delta_{ij}(t^{2H} + s^{2H} - |t - s|^{2H}).$$

That is, the components of B^H are independent fractional Brownian motions with Hurst parameter H (see Example 3.5.3).

The self-intersection local time of B^H on $[0, T]$ is heuristically defined by

$$I_T = \int_0^T \int_0^t \delta_0(B_t^H - B_s^H)\,ds\,dt.$$

Notice that

$$E(I_T) = \int_0^T \int_0^t E(\delta_0(B_t^H - B_s^H))\,ds\,dt$$

$$= (2\pi)^{-d/2} \int_0^T \int_0^t |t - s|^{-Hd}\,ds\,dt < \infty \quad \Longleftrightarrow \quad Hd < 1.$$

This informal computation predicts that the self-intersection local time will exist only if $Hd < 1$. In the Brownian motion case (that is, $H = 1/2$) this means $d = 1$.

A rigorous definition of I_T is obtained by taking the limit in $L^2(\Omega)$ as $\epsilon \to 0$ of the approximation

$$I_{T,\epsilon} = \int_0^T \int_0^t p_\epsilon(B_t^H - B_s^H)\,ds\,dt,$$

where $p_\epsilon(x) = (2\pi\epsilon)^{-d/2} \exp(-|x|^2/(2\epsilon))$. Then, it can be proved that $I_{T,\epsilon}$ converges in $L^2(\Omega)$ if and only if $Hd < 1$. When $Hd > 1$, we introduce the so-called *Varadhan renormalization*, which consists of subtracting the expectation $E(I_{T,\epsilon})$; the latter is the divergent term. Then, for $1/d \leq H < 3/(2d)$,

$$I_{T,\epsilon} - E(I_{T,\epsilon}) \xrightarrow{L^2(\Omega)} \tilde{I}_T,$$

and $\tilde{I}_T$ is called the renormalized self-intersection local time (see Hu and Nualart, 2005). For Brownian motion (that is, $H = 1/2$) this happens when $d = 2$.

When $H \geq 3/(2d)$, with proper normalization we can establish a central limit theorem.

Theorem 6.6.1 (Hu and Nualart, 2005)   *Let B^H be a d-dimensional fractional Brownian motion with Hurst parameter* H. *Assume that $d \geq 2$. Then, if $3/(2d) < H < 3/4$, we have the convergence in law*

$$\epsilon^{d/2-3/(4H)} (I_{T,\epsilon} - E(I_{T,\epsilon})) \xrightarrow{\mathcal{L}} N(0, T\sigma_{H,d}^2) \qquad (6.12)$$

as $\epsilon \downarrow 0$.

The case $H = 3/(2d)$ was addressed as well in Hu and Nualart (2005), where it was shown that the sequence $(\log(1/\epsilon))^{-1/2}(I_T^\epsilon - E(I_T^\epsilon))$ converges in law to a centered Gaussian distribution with variance $T\sigma_{log}^2$ as $\epsilon \to 0$, where the constant σ_{log}^2 is given by equation (42) in Hu and Nualart (2005).

Theorem 6.6.1 was proved in Hu and Nualart (2005) using the Wiener chaos expansion of $I_{T,\epsilon}$ and applying Theorem 8.4.1 below. The variance limit is given by

$$\sigma_{H,d}^2 = (2\pi)^{-d} \int_{\substack{0<s<t<T \\ 0<s'<t'<T}} ((\lambda\rho - \mu^2)^{-d/2} - (\lambda\rho)^{-d/2})ds\,dt\,ds'\,dt', \qquad (6.13)$$

with the notation $\lambda = |t - s|^{2H}$, $\rho = |t' - s'|^{2H}$, and $\mu = E((B_t^{H,1} - B_s^{H,1})(B_{t'}^{H,1} - B_{s'}^{H,1}))$. To show that $\sigma_{H,d}^2 < \infty$ we need to estimate the double integral appearing in (6.13) over essentially three types of region: $[s', t'] \subset [s, t]$, $[s', t'] \cap [s, t] = \emptyset$, and $s < s' < t < t'$ (the intervals overlap).

A natural extension of Theorem 6.6.1 is the following functional version, established in Jaramillo and Nualart (2018).

Theorem 6.6.2   *Let B^H be a d-dimensional fractional Brownian motion with Hurst parameter* H. *Assume that $d \geq 2$ and $3/(2d) < H < 3/4$. Then*

$$\left(\epsilon^{d/2-3/(4H)}(I_{T,\epsilon} - E(I_{T,\epsilon}))\right)_{T\geq 0} \xrightarrow{\mathcal{L}} (\sigma_{H,d}B_T)_{T\geq 0}, \qquad (6.14)$$

in the space $C([0, \infty))$ endowed with the topology of uniform convergence on compact sets, where B is a standard Brownian motion.

The proof of the convergence of the finite-dimensional distributions follows using the same approach as in Theorem 6.6.1. Then, the main difficulty in establishing the convergence (6.14) is to show the tightness property of the laws. From the Billingsley criterion (see Billingsley, 1999, Theorem 12.3), the tightness property can be derived by showing that there exists $p > 2$ such that, for every $0 \leq S \leq T$,

$$E(|\Phi_{T,\epsilon} - \Phi_{S,\epsilon}|^p) \leq C_{p,d,H}|T - S|^{p/2}, \qquad (6.15)$$

for some constant $C > 0$ independent of S, T and ϵ, where

$$\Phi_{T,\epsilon} = \epsilon^{d/2 - 3/(4H)}(I_{T,\epsilon} - E(I_{T,\epsilon})). \qquad (6.16)$$

The problem in finding a bound of the type (6.15) comes from the fact that the smallest even integer $p > 2$ is $p = 4$, and, in view of the above comments for $p = 2$, a direct computation of the moment of order four, $E(|\Phi_{T,\epsilon} - \Phi_{S,\epsilon}|^4)$, is too complicated to be handled. To overcome this difficulty, a new approach to prove tightness based on the techniques of Malliavin calculus has been developed in Jaramillo and Nualart (2018). We describe the main ingredients of this approach in the next proposition.

Proposition 6.6.3 *With the above notation, for all $2 < p < 4Hd/3$ we have, for all $0 \leq S \leq T$,*

$$E(|\Phi_{T,\epsilon} - \Phi_{S,\epsilon}|^p) \leq C_{p,d,H}|T - S|^{p/2},$$

Proof Fix $S \leq T$ and define $Z_\epsilon = \Phi_{T,\epsilon} - \Phi_{S,\epsilon}$, where $\Phi_{T,\epsilon}$ was defined in (6.16). The representation established in Proposition 6.5.1 applied to the centered random variable Z_ϵ yields

$$Z_\epsilon = -\delta DL^{-1}Z_\epsilon.$$

Using that $E(DL^{-1}Z_\epsilon) = 0$ and Meyer's inequality (see (5.7)), we obtain

$$\|Z_\epsilon\|_p \leq c_p\|D^2L^{-1}Z_\epsilon\|_{L^p(\Omega;(H^d)^{\otimes 2})},$$

where H is the Hilbert space associated with the covariance of the fractional Brownian motion. We know that on the one hand

$$Z_\epsilon = \epsilon^{d/2 - 3/(4H)} \int_{\substack{s < t \\ S < t < T}} \left(p_\epsilon(B_t^H - B_s^H) - E\left(p_\epsilon(B_t^H - B_s^H)\right) \right) ds\, dt.$$

On the other hand, for any centered random variable F we have

$$L^{-1}F = -\int_0^\infty T_\theta F\, d\theta,$$

where $(T_\theta)_{\theta \geq 0}$ is the Ornstein–Uhlenbeck semigroup (see Exercise 5.5). As a consequence, from Mehler's formula (5.1) we can write

$$D^2 L^{-1} Z_\epsilon = -\int_0^\infty D^2 T_\theta Z_\epsilon d\theta$$

$$= -\int_0^\infty \int_{\substack{s<t \\ S<t<T}} D^2 T_\theta (p_\epsilon(B_t^H - B_s^H)) ds dt d\theta$$

$$= -\int_0^\infty \int_{\substack{s<t \\ S<t<T}} D^2 \tilde{E} \left(p_\epsilon (e^{-\theta}(B_t^H - B_s^H) + \sqrt{1 - e^{-2\theta}}(\tilde{B}_t^H - \tilde{B}_s^H)) \right) ds dt d\theta$$

$$= -\int_0^\infty \int_{\substack{s<t \\ S<t<T}} D^2 p_{\epsilon+(1-e^{-2\theta})(t-s)^{2H}} (e^{-\theta}(B_t^H - B_s^H)) ds dt d\theta$$

$$= -\int_0^\infty \int_{\substack{s<t \\ S<t<T}} e^{-2\theta} p''_{\epsilon+(1-e^{-2\theta})(t-s)^{2H}} (e^{-\theta}(B_t^H - B_s^H)) \mathbf{1}_{[s,t]}^{\otimes 2} ds dt d\theta,$$

where p'' denotes the Hessian matrix of p. Finally, using Minkowski's inequality we obtain

$$\|Z_\epsilon\|_p^2 \leq c_p^2 \|D^2 L^{-1} Z_\epsilon\|_{L^p(\Omega;(H^d)^{\otimes 2})}^2 = c_p^2 \|\|DL^{-1}Z_\epsilon\|_{(H^d)^{\otimes 2}}^2\|_{p/2}$$

$$\leq c_p^2 \int_{\mathbb{R}_+^2} \int_{\substack{s<t \\ S<t<T}} \int_{\substack{s'<t' \\ S<t'<T}} e^{-2\theta-2\beta} \mu^2 \|p''_{\epsilon+(1-e^{-2\theta})}(e^{-\theta}(B_t^H - B_s^H))\|_p$$

$$\times \|p''_{\epsilon+(1-e^{-2\theta})}(e^{-\beta}(B_t^H - B_s^H))\|_p ds dt ds' dt' d\theta d\beta.$$

Performing some Gaussian computations (see Jaramillo and Nualart, 2018, for the details) we arrive at the estimate

$$\|Z_\epsilon\|_p^2 \leq K_{p,d,H} |T - S|^p,$$

where

$$K_{p.d.H} = \int_{\substack{0<s<t<T \\ 0<s'<t'<T}} \frac{\mu^2}{\lambda \rho} ((1 + \lambda)(1 + \rho) - \mu^2)^{-d/p} ds dt ds' dt' < \infty,$$

where $\lambda, \rho,$ and μ are the functions of s, t, s', t' appearing in the double integral (6.13) and K is a constant depending on p, d, H. Then, using arguments similar to those in the proof of $\sigma_{H,d}^2 < \infty$, one can show that $K_{p,d,H} < \infty$ provided that $2 < p < 4Hd/3$ (see Jaramillo and Nualart, 2018, for the details). This completes the proof. $\square$

Exercises

6.1 Using the Clark–Ocone formula find the stochastic integral representation of the random variables in Exercise 2.13.

6.2   Let $F = I_n(f)$, $f \in L_s^2([0,T]^n)$, and for any $A \subset \mathcal{B}(\mathbb{R})$ consider the σ-field $\mathcal{F}_A = \sigma\{B(1_D), D \subset A\}$. Show that

$$E(F|\mathcal{F}_A) = I_n(f_n 1_A^{\otimes n}).$$

6.3   Let $F = I_n(f)$, $f \in L_s^2([0,T]^n)$. Use the Clark–Ocone formula to show that $F = \delta(u)$, where $u_t = nI_{n-1}(f(\cdot,t)1_{[0,t]})$. Deduce that $u_t = E(D_t F|\mathcal{F}_t)$. Apply this approach to prove the Clark–Ocone formula via the Wiener chaos expansion.

6.4   Let $F \in L^2(\Omega, \mathcal{F}_{[a,b]}, P)$. Show that F admits the following representation:

$$F = E(F|\mathcal{F}_{[a,t]^c}) + \int_a^b E(D_t F|\mathcal{F}_{[a,t]^c})dB_t.$$

7

Study of Densities

In this chapter we apply Malliavin calculus to derive explicit formulas for the densities of random variables on Wiener space and to establish criteria for their regularity. We apply these criteria to the proof of Hörmander's hypoellipticity theorem.

7.1 Analysis of Densities in the One-Dimensional Case

We recall that $B = (B_t)_{t \geq 0}$ is a Brownian motion on a probability space $(\Omega, \mathcal{F}, P)$ such that $\mathcal{F}$ is generated by B. The topological support of the law of a random variable F is defined as the set of points $x \in \mathbb{R}$ such that $P(|x - F| < \epsilon) > 0$ for all $\epsilon > 0$.

Our first result says that if a random variable F belongs to the Sobolev space $\mathbb{D}^{1,2}$ then the topological support of the law of F is a closed interval.

Proposition 7.1.1 *Let $F \in \mathbb{D}^{1,2}$. Then, the topological support of the law of F is a closed interval.*

Proof Clearly the topological support of the law of F is a closed set. Then, it suffices to show that it is connected. We show this by contradiction. If the topological support of the law of F is not connected, there exists a point $a \in \mathbb{R}$ and $\epsilon > 0$ such that $P(a - \epsilon < F < a + \epsilon) = 0$, $P(F \geq a + \epsilon) < 1$, and $P(F \leq a - \epsilon) < 1$. Let $\varphi : \mathbb{R} \to \mathbb{R}$ be an infinitely differentiable function such that $\varphi(x) = 0$ if $x \leq a - \epsilon$ and $\varphi(x) = 1$ if $x \geq a + \epsilon$. By Proposition 3.3.2, $\varphi(F) \in \mathbb{D}^{1,2}$ but, almost surely, $\varphi(F) = \mathbf{1}_{\{F \geq a + \epsilon\}}$. Therefore, by Proposition 4.2.6, we must have $P(F \geq a + \epsilon) = 0$ or $P(F \geq a + \epsilon) = 1$, which leads to a contradiction. □

If a random variable F belongs to $\mathbb{D}^{1,2}$, and its derivative is not degenerate, then F has a density. A simple proof of this result was given by Nualart and Zakai (1986).

Proposition 7.1.2 *Let F be a random variable in the space $\mathbb{D}^{1,2}$ such that*

$\|DF\|_H > 0$ *almost surely. Then, the law of F is absolutely continuous with respect to the Lebesgue measure on* $\mathbb{R}$.

Proof Replacing F by arctan F, we may assume that F takes values in $(-1, 1)$. It suffices to show that, for any measurable function $g : (-1, 1) \to [0, 1]$ such that $\int_{-1}^{1} g(y)dy = 0$, we have $\mathrm{E}(g(F)) = 0$. We can find a sequence of continuous functions $g_n : (-1, 1) \to [0, 1]$ such that, as n tends to infinity, $g_n(y)$ converges to $g(y)$ for almost all y with respect to the measure $\mathrm{P} \circ F^{-1} + \ell$, where ℓ denotes the Lebesgue measure on $\mathbb{R}$. Set

$$\psi_n(x) = \int_{-\infty}^{x} g_n(y)dy.$$

Then, $\psi_n(F)$ converges to 0 almost surely and in $L^2(\Omega)$ because g_n converges almost everywhere to g, with respect to the Lebesgue measure, and $\int_{-1}^{1} g(y)dy = 0$. Furthermore, by the chain rule (Proposition 3.3.2), $\psi_n(F) \in \mathbb{D}^{1,2}$ and

$$D(\psi_n(F)) = g_n(F)DF,$$

which converges almost surely and in $L^2(\Omega)$ to $g(F)DF$. Because D is closed, we conclude that $g(F)DF = 0$. Our hypothesis $\|DF\|_H > 0$ implies that $g(F) = 0$ almost surely, and this finishes the proof. $\square$

The following result is an expression for the density of a random variable in the Sobolev space $\mathbb{D}^{1,2}$, assuming that $\|DF\|_H > 0$ a.s.

Proposition 7.1.3 *Let F be a random variable in the space* $\mathbb{D}^{1,2}$ *such that* $\|DF\|_H > 0$ *a.s. Suppose that* $DF/\|DF\|_H^2$ *belongs to the domain of the operator* δ *in* $L^2(\Omega)$. *Then the law of F has a continuous and bounded density, given by*

$$p(x) = \mathrm{E}\left(\mathbf{1}_{\{F>x\}}\delta\left(\frac{DF}{\|DF\|_H^2}\right)\right). \tag{7.1}$$

Proof Let ψ be a nonnegative function in $C_0^\infty(\mathbb{R})$, and set $\varphi(y) = \int_{-\infty}^{y} \psi(z)dz$. Then, by the chain rule (Proposition 3.3.2), $\varphi(F)$ belongs to $\mathbb{D}^{1,2}$ and we can write

$$\langle D(\varphi(F)), DF\rangle_H = \psi(F)\|DF\|_H^2.$$

Using the duality formula (Proposition 3.2.3), we obtain

$$\mathrm{E}(\psi(F)) = \mathrm{E}\left(\left\langle D(\varphi(F)), \frac{DF}{\|DF\|_H^2}\right\rangle_H\right) = \mathrm{E}\left(\varphi(F)\delta\left(\frac{DF}{\|DF\|_H^2}\right)\right). \tag{7.2}$$

By an approximation argument, equation (7.2) holds for $\psi(y) = \mathbf{1}_{[a,b]}(y)$, where $a < b$. As a consequence, we can apply Fubini's theorem to get

$$P(a \le F \le b) = E\left(\left(\int_{-\infty}^{F} \psi(x)dx\right)\delta\left(\frac{DF}{\|DF\|_H^2}\right)\right)$$

$$= \int_{a}^{b} E\left(\mathbf{1}_{\{F>x\}}\delta\left(\frac{DF}{\|DF\|_H^2}\right)\right)dx,$$

which implies the desired result.   □

Remark 7.1.4   Equation (7.1) still holds under the hypotheses $F \in \mathbb{D}^{1,p}$ and $DF/\|DF\|_H^2 \in \mathbb{D}^{1,p'}(H)$ for some $p, p' > 1$. Sufficient conditions for these hypotheses are $F \in \mathbb{D}^{2,\alpha}$ and $E(\|DF\|^{-2\beta}) < \infty$ with $1/\alpha + 1/\beta < 1$.

Example 7.1.5   Let $F = B(h)$. Then $DF = h$ and

$$\delta\left(\frac{DF}{\|DF\|_H^2}\right) = B(h)\|h\|_H^{-2}.$$

As a consequence, formula (7.1) yields

$$p(x) = \|h\|_H^{-2} E\left(\mathbf{1}_{\{F>x\}}F\right),$$

which is true because $p(x)$ is the density of the distribution $N(0, \|h\|_H^2)$ (Exercise 7.1).

Applying equation (7.1) we can derive density estimates. Notice first that (7.1) holds if $\mathbf{1}_{\{F>x\}}$ is replaced by $\mathbf{1}_{\{F<x\}}$, because the divergence has zero expectation. Fix p and q such that $1/p + 1/q = 1$. Then, by Hölder's inequality, we obtain

$$p(x) \le (P(|F| > |x|))^{1/q}\left\|\delta\left(\frac{DF}{\|DF\|_H^2}\right)\right\|_p, \qquad (7.3)$$

for all $x \in \mathbb{R}$. Applying (7.3) and Meyer's inequality (5.7), we can deduce the following result (see Nualart, 2006, Proposition 2.1.3). The proof is left as an exercise (Exercise 7.2).

Proposition 7.1.6   *Let q, α, β be three positive real numbers such that $1/q + 1/\alpha + 1/\beta = 1$. Let F be a random variable in the space $\mathbb{D}^{2,\alpha}$, such that $E(\|DF\|_H^{-2\beta}) < \infty$. Then, the density $p(x)$ of F can be estimated as follows:*

$$p(x) \le c_{q,\alpha,\beta}\,(P(|F| > |x|))^{1/q}$$

$$\times\left(E(\|DF\|_H^{-1}) + \|D^2F\|_{L^\alpha(\Omega;L^2(\mathbb{R}_+^2))}\,\|\|DF\|_H^{-2}\|_\beta\right). \qquad (7.4)$$

7.2 Existence and Smoothness of Densities for Random Vectors

Let $F = (F^1, \ldots, F^m)$ be such that $F^i \in \mathbb{D}^{1,2}$ for $i = 1, \ldots, m$. We define the *Malliavin matrix* of F as the random symmetric nonnegative definite matrix

$$\gamma_F = (\langle DF^i, DF^j \rangle_H)_{1 \le i,j \le m}. \tag{7.5}$$

In the one-dimensional case, $\gamma_F = \|DF\|_H^2$. The following theorem is a multidimensional version of Proposition 7.1.2.

Theorem 7.2.1 *If* $\det \gamma_F > 0$ *a.s. then the law of F is absolutely continuous with respect to the Lebesgue measure on* $\mathbb{R}^m$.

This theorem was proved by Bouleau and Hirsch (1991) using the coarea formula and techniques of geometric measure theory, and we omit the proof. As a consequence, the measure $(\det \gamma_F \times P) \circ F^{-1}$ is always absolutely continuous; that is,

$$P(F \in B, \det \gamma_F > 0) = 0,$$

for any Borel set $B \in \mathcal{B}(\mathbb{R}^m)$ of zero Lebesgue measure.

Definition 7.2.2 We say that a random vector $F = (F^1, \ldots, F^m)$ is *non-degenerate* if $F^i \in \mathbb{D}^{1,2}$ for $i = 1, \ldots, m$ and

$$E((\det \gamma_F)^{-p}) < \infty,$$

for all $p \ge 2$.

Set $\partial_i = \partial/\partial x_i$ and, for any multi-index $\alpha \in \{1, \ldots, m\}^k$, $k \ge 1$, we denote by ∂_α the partial derivative $\partial^k / (\partial x_{\alpha_1} \cdots \partial x_{\alpha_k})$.

Lemma 7.2.3 *Let γ be an $m \times m$ random matrix such that $\gamma^{ij} \in \mathbb{D}^{1,\infty}$ for all i, j and $E(|\det \gamma|^{-p}) < \infty$ for all $p \ge 2$. Then, $(\gamma^{-1})^{ij}$ belongs to $\mathbb{D}^{1,\infty}$ for all i, j, and*

$$D(\gamma^{-1})^{ij} = -\sum_{k,\ell=1}^{m} (\gamma^{-1})^{ik} (\gamma^{-1})^{\ell j} D\gamma^{k\ell}. \tag{7.6}$$

Proof We know that $P(\det \gamma > 0)$ is zero or one (see Exercise 4.10). So, we can assume that $\det \gamma > 0$ almost surely. For any $\epsilon > 0$, we define $\gamma_\epsilon^{-1} = (\det \gamma + \epsilon)^{-1} A(\gamma)$, where $A(\gamma)$ is the adjoint matrix of γ. Then, the entries of γ_ϵ^{-1} belong to $\mathbb{D}^{1,\infty}$ and converge in $L^p(\Omega)$, for all $p \ge 2$, to those of γ^{-1} as ϵ tends to zero. Moreover, the entries of γ_ϵ^{-1} satisfy

$$\sup_{\epsilon \in (0,1]} \|(\gamma_\epsilon^{-1})^{ij}\|_{1,p} < \infty,$$

for all $p \geq 2$. Therefore, by Proposition 5.1.6 the entries of γ_ϵ^{-1} belong to $\mathbb{D}^{1,p}$ for any $p \geq 2$. Finally, from the expression $\gamma_\epsilon^{-1}\gamma = (\det\gamma/(\det\gamma + \epsilon))I_m$, where I_m denotes the identity matrix of order m, we deduce (7.6) on applying the derivative operator and letting ϵ tend to zero. $\qquad\square$

The following result can be regarded as an integration-by-parts formula and plays a fundamental role in the proof of the regularity of densities.

Proposition 7.2.4 *Let $F = (F^1, \ldots, F^m)$ be a nondegenerate random vector. Fix $k \geq 1$ and suppose that $F^i \in \mathbb{D}^{k+1,\infty}$ for $i = 1, \ldots, m$. Let $G \in \mathbb{D}^\infty$ and let $\varphi \in C_p^\infty(\mathbb{R}^m)$. Then, for any multi-index $\alpha \in \{1, \ldots, m\}^k$, there exists an element $H_\alpha(F, G) \in \mathbb{D}^\infty$ such that*

$$E(\partial_\alpha\varphi(F)G) = E(\varphi(F)H_\alpha(F, G)), \tag{7.7}$$

where the elements $H_\alpha(F, G)$ are recursively given by

$$H_{(i)}(F, G) = \sum_{j=1}^m \delta\left(G\left(\gamma_F^{-1}\right)^{ij} DF^j\right)$$

and, for $\alpha = (\alpha_1, \ldots, \alpha_k)$, $k \geq 2$, we set

$$H_\alpha(F, G) = H_{\alpha_k}(F, H_{(\alpha_1, \ldots, \alpha_{k-1})}(F, G)).$$

Proof By the chain rule (Proposition 3.3.2), we have

$$\langle D(\varphi(F)), DF^j\rangle_H = \sum_{i=1}^m \partial_i\varphi(F)\langle DF^i, DF^j\rangle_H = \sum_{i=1}^m \partial_i\varphi(F)\gamma_F^{ij}$$

and, consequently,

$$\partial_i\varphi(F) = \sum_{j=1}^m \langle D(\varphi(F)), DF^j\rangle_H(\gamma_F^{-1})^{ji}.$$

Taking expectations and using the duality relationship (Proposition 3.2.3) yields

$$E(\partial_i\varphi(F)G) = E(\varphi(F)H_{(i)}(F, G)),$$

where $H_{(i)} = \sum_{j=1}^m \delta\left(G\left(\gamma_F^{-1}\right)^{ij} DF^j\right)$. Notice that Meyer's inequality (Theorem 5.3.1) and Lemma 7.2.3 imply that $H_{(i)}$ belongs to $L^p(\Omega)$ for any $p \geq 2$. We finish the proof with a recurrence argument. $\qquad\square$

One can show that, for any $p > 1$, there exist constants β, $\gamma > 1$ and integers n, m such that

$$\|H_\alpha(F, G)\|_p \leq c_{p,q}\left\|\det\gamma_F^{-1}\right\|_\beta^m \|DF\|_{k,\gamma}^n \|G\|_{k,q}. \tag{7.8}$$

The proof of this inequality is based on Meyer's and Hölder's inequalities and it is left as an exercise (Exercise 7.4).

The following result is a multidimensional version of the density formula (7.1).

Proposition 7.2.5 *Let $F = (F^1, \ldots, F^m)$ be a nondegenerate random vector such that $F^i \in \mathbb{D}^{m+1,\infty}$ for $i = 1, \ldots, m$. Then F has a continuous and bounded density given by*

$$p(x) = E(\mathbf{1}_{\{F>x\}} H_\alpha(F, 1)), \tag{7.9}$$

where $\alpha = (1, 2, \ldots, m)$.

Proof   Recall that, for $\alpha = (1, 2, \ldots, m)$

$$H_\alpha(F, 1)$$
$$= \sum_{j_1, \ldots, j_m = 1}^{m} \delta\left((\gamma_F^{-1})^{1 j_1} DF^{j_1} \delta\left((\gamma_F^{-1})^{2 j_2} DF^{j_2} \cdots \delta\left((\gamma_F^{-1})^{m j_m} DF^{j_m}\right) \cdots \right)\right).$$

Then, equality (7.7) applied to the multi-index $\alpha = (1, 2, \ldots, m)$ yields, for any $\varphi \in C_p^\infty(\mathbb{R}^m)$,

$$E(\partial_\alpha \varphi(F)) = E(\varphi(F) H_\alpha(F, 1)).$$

Notice that

$$\varphi(F) = \int_{-\infty}^{F^1} \cdots \int_{-\infty}^{F^m} \partial_\alpha \varphi(x) dx.$$

Hence, by Fubini's theorem we can write

$$E(\partial_\alpha \varphi(F)) = \int_{\mathbb{R}^m} \partial_\alpha \varphi(x) E(\mathbf{1}_{\{F>x\}} H_\alpha(F, 1)) dx. \tag{7.10}$$

Given any function $\psi \in C_0^\infty(\mathbb{R}^m)$, we can take $\varphi \in C_p^\infty(\mathbb{R}^m)$ such that $\psi = \partial_\alpha \varphi$, and (7.10) yields

$$E(\psi(F)) = \int_{\mathbb{R}^m} \psi(x) E(\mathbf{1}_{\{F>x\}} H_\alpha(F, 1)) dx,$$

which implies the result. $\square$

The following theorem is the basic criterion for the smoothness of densities.

Theorem 7.2.6 *Let $F = (F^1, \ldots, F^m)$ be a nondegenerate random vector such that $F^i \in \mathbb{D}^\infty$ for all $i = 1, \ldots, m$. Then the law of F possesses an infinitely differentiable density.*

Proof For any multi-index β and any $\varphi \in C_p^\infty(\mathbb{R}^m)$, we have, taking $\alpha = (1, 2, \ldots, m)$,

$$E\left(\partial_\beta \partial_\alpha \varphi(F)\right) = E\left(\varphi(F) H_\beta(F, H_\alpha(F, 1))\right)$$

$$= \int_{\mathbb{R}^m} \partial_\alpha \varphi(x) E\left(1_{\{F > x\}} H_\beta(F, H_\alpha(F, 1))\right) dx.$$

Hence, for any $\xi \in C_0^\infty(\mathbb{R}^m)$,

$$\int_{\mathbb{R}^m} \partial_\beta \xi(x) p(x) dx = \int_{\mathbb{R}^m} \xi(x) E\left(1_{\{F > x\}} H_\beta(F, H_\alpha(F, 1))\right) dx.$$

Therefore, $p(x)$ is infinitely differentiable and, for any multi-index β, we have

$$\partial_\beta p(x) = (-1)^{|\beta|} E\left(1_{\{F > x\}} H_\beta(F, (H_\alpha(F, 1))\right).$$

This completes the proof. $\qquad\qquad\qquad\qquad\qquad\qquad\qquad\qquad\qquad$ $\square$

7.3 Density Formula using the Riesz Transform

In this section we present a method for obtaining a density formula using the Riesz transform, following the methodology introduced by Malliavin and Thalmaier (2005) and extensively studied by Bally and Caramellino (2011, 2013). In contrast with (7.9), here we only need two derivatives, instead of $m + 1$.

Let Q_m be the fundamental solution to the Laplace equation $\Delta Q_m = \delta_0$ on $\mathbb{R}^m$, $m \geq 2$. That is,

$$Q_2(x) = a_2^{-1} \ln \frac{1}{|x|}, \quad Q_m(x) = a_m^{-1} |x|^{2-m}, \quad m > 2,$$

where a_m is the area of the unit sphere in $\mathbb{R}^m$. We know that, for any $1 \leq i \leq m$,

$$\partial_i Q_m(x) = -c_m \frac{x_i}{|x|^m}, \tag{7.11}$$

where $c_m = 2(m-2)/a_m$ if $m > 2$ and $c_2 = 2/a_2$. Notice that any function φ in $C_0^1(\mathbb{R}^m)$ can be written as

$$\varphi(x) = \nabla \varphi * \nabla Q_m(x) = \sum_{i=1}^m \int_{\mathbb{R}^m} \partial \varphi(x - y) \partial_i Q_m(y) dy. \tag{7.12}$$

Indeed,

$$\nabla \varphi * \nabla Q_m(x) = \varphi * \Delta Q_m(x) = \varphi(x).$$

Theorem 7.3.1 *Let F be an m-dimensional nondegenerate random vector whose components are in* $\mathbb{D}^{2,\infty}$. *Then, the law of F admits a continuous and bounded density p given by*

$$p(x) = \sum_{i=1}^{m} \mathrm{E}(\partial_i Q_m(F - x)H_{(i)}(F, 1)),$$

where

$$H_{(i)}(F, 1) = \sum_{j=1}^{m} \delta\left((\gamma_F^{-1})^{ij}DF^j\right).$$

Proof Let $\varphi \in C_0^1(\mathbb{R}^m)$. Applying (7.12), we can write

$$\mathrm{E}(\varphi(F)) = \sum_{i=1}^{m} \mathrm{E}\left(\int_{\mathbb{R}^m} \partial_i Q_m(y)(\partial_i\varphi(F - y))dy\right).$$

Assume that the support of φ is included in the ball $B_R(0)$ for some $R > 1$. Then, using (7.11) we obtain

$$\mathrm{E}\left(\int_{\mathbb{R}^m} |\partial_i Q_m(y)\partial_i\varphi(F - y)|\, dy\right) \leq \|\partial_i\varphi\|_\infty \mathrm{E}\left(\int_{\{y:|F|-R\leq|y|\leq|F|+R\}} |\partial_i Q_m(y)|dy\right)$$

$$\leq c_m \mathrm{Vol}(B_1(0))\|\partial_i\varphi\|_\infty \mathrm{E}\left(\int_{|F|-R}^{|F|+R} \frac{r}{r^m}r^{m-1}dr\right)$$

$$= 2c_m \mathrm{Vol}(B_1(0))\|\partial_i\varphi\|_\infty R\mathrm{E}(|F|) < \infty.$$

As a consequence, Fubini's theorem and (7.7) yield

$$\mathrm{E}(\varphi(F)) = \sum_{i=1}^{m} \int_{\mathbb{R}^m} \partial_i Q_m(y)\mathrm{E}(\partial_i\varphi(F - y))dy$$

$$= \sum_{i=1}^{m} \int_{\mathbb{R}^m} \partial_i Q_m(y)\mathrm{E}(\varphi(F - y)H_{(i)}(F, 1))dy$$

$$= \sum_{i=1}^{m} \int_{\mathbb{R}^m} \varphi(y)\mathrm{E}(\partial_i Q_m(F - y)H_{(i)}(F, 1))dy.$$

This completes the proof.   □

The approach based on the Riesz transform can also be used to obtain the following uniform estimate for densities, due to Stroock.

Lemma 7.3.2 *Under the assumptions of Theorem 7.3.1, for any p > m there exists a constant c depending only on m and p such that*

$$\|p\|_\infty \leq c\left(\max_{1\leq i\leq m} \|H_{(i)}(F, 1)\|_p\right)^m.$$

Proof From

$$p(x) = \sum_{i=1}^{m} \mathrm{E}(\partial_i Q_m(F - x) H_{(i)}(F, 1)),$$

applying Hölder's inequality with $1/p + 1/q = 1$ and the estimate (see (7.11))

$$|\partial_i Q_m(F - x)| \le c_m |F - x|^{1-m}$$

yields

$$p(x) \le m c_m A \left(\mathrm{E}\left(|F - x|^{(1-m)q} \right) \right)^{1/q}, \tag{7.13}$$

where $A = \max_{1 \le i \le m} \|H_{(i)}(F, 1)\|_p$.

Suppose first that p is bounded and let $M = \sup_{x \in \mathbb{R}} p(x)$. We can write, for any $\epsilon > 0$,

$$\mathrm{E}(|F - x|^{(1-m)q}) \le \epsilon^{(1-m)q} + \int_{|z-x| \le \epsilon} |z - x|^{(1-m)q} p(x) dx$$

$$\le \epsilon^{(1-m)q} + C_{m,p} \epsilon^{(p-m)/(p-1)} M. \tag{7.14}$$

Therefore, substituting (7.14) into (7.13), we get

$$M \le A m c_m \left(\epsilon^{1-m} + C_{m,p}^{1/q} \epsilon^{(p-m)/p} M^{1/q} \right).$$

Now we minimize with respect to ϵ and obtain $M \le A C_{m,p} M^{1-1/m}$, for some constant $C_{m,p}$, which implies that $M \le C_{m,p}^m A^m$. If p is not bounded, we apply the procedure to $p * \psi_\delta$, where ψ_δ is an approximation of the identity, and let δ tend to zero at the end. $\square$

7.4 Log-Likelihood Density Formula

The next result gives an expression for the derivative of the logarithm of the density of a random vector. It is inspired by the concept of a covering vector field introduced in Malliavin and Thalmaier (2005, Definition 4.4).

Definition 7.4.1 Let F be an m-dimensional random vector whose components are in $\mathbb{D}^{1,2}$. An m-dimensional process $u = (u_k(t))_{t \ge 0, 1 \le k \le m}$ is called a *covering vector field* of F if, for any $k = 1, \ldots, m$, $u_k \in \mathrm{Dom}\, \delta$ and

$$\partial_k \varphi(F) = \langle D(\varphi(F)), u_k \rangle_H$$

for any $\varphi \in C_0^1(\mathbb{R}^m)$.

For instance, $u = \gamma_F DF$ is a covering vector field of F, where γ_F is the Malliavin matrix of F introduced in (7.5). Observe that, by the duality relationship (Proposition 3.2.3), if u is a covering vector field of F then the following integration-by-parts formula holds:

$$E(\partial_k \varphi(F)) = E(\varphi(F)\delta(u_k)). \tag{7.15}$$

Moreover, we have an expression for the derivative of the logarithm of the density.

Proposition 7.4.2 *Consider an m-dimensional nondegenerate random vector F whose components are in $\mathbb{D}^\infty$. Suppose that $p(x) > 0$ a.e., where p denotes the density of F. Then, for any covering vector field u and all $k \in \{1, \ldots, m\}$,*

$$\partial_k \log p(y) = -E(\delta(u_k)|F = y) \quad a.e. \tag{7.16}$$

Proof First observe that by Theorem 7.2.6 p is in $C^\infty(\mathbb{R}^m)$. Using (7.15) we get, for any $\varphi \in C_0^\infty(\mathbb{R}^m)$,

$$\int_{\mathbb{R}^m} \varphi(y)\partial_k p(y)dy = -\int_{\mathbb{R}^m} \varphi(y)E(\delta(u_k)|F = y)p(y)dy,$$

and the desired result follows. $\qquad\square$

We next derive a formula for the log-likelihood function of a statistical observation in a parametric statistical model using the techniques of Malliavin calculus. This application of the Malliavin calculus to parametric statistical models was initiated by Gobet (2001, 2002) in order to obtain asymptotic results for parametric diffusion models. See also Corcuera and Kohatsu-Higa (2011).

Let $\Theta \subset \mathbb{R}$ be an open interval. A statistical observation is described by an n-dimensional random vector $F_\theta = (F_{\theta,1}, \ldots, F_{\theta,n})$, depending on an unknown parameter $\theta \in \Theta$ defined on a probability space $(\Omega, \mathcal{F}, P)$. We denote by P_θ the probability distribution of F_θ and by E_θ the expectation with respect to P_θ. We call $(F_\theta)_{\theta \in \Theta}$ a parametric statistical model.

We are going to impose the existence of a log-likelihood density function for P_θ that satisfies some smoothness conditions. These conditions are included in the notion of a regular model (see Ibragimov and Has'minskii (1981) for an extended exposition of statistical estimation).

Definition 7.4.3 A parametric statistical model $(F_\theta)_{\theta \in \Theta}$ is said to be *regular* if:

(i) There exists a σ-finite measure μ on $\mathbb{R}^n$ such that, for all $\theta \in \Theta$, the probability measures $P_\theta = P \circ F_\theta^{-1}$ are absolutely continuous with respect to μ, and the density

$$p(x, \theta) = \frac{dP_\theta}{d\mu}(x) \qquad (7.17)$$

is positive and continuous in Θ for μ-almost-all $x \in \mathbb{R}^n$. The measure μ is called the dominating measure of the model.

(ii) The density $p(x, \theta)$ is differentiable in Θ for μ-almost-all $x \in \mathbb{R}^n$, and the derivative of $p^{1/2}(x, \theta)$ belongs to $L^2(\mu)$. That is, the function

$$\varphi(x, \theta) = \partial_\theta p^{1/2}(x, \theta) = \frac{1}{2} \frac{1}{p^{1/2}(x, \theta)} \partial_\theta p(x, \theta)$$

satisfies that, for all $\theta \in \Theta$,

$$\int_{\mathbb{R}^n} |\varphi(x, \theta)|^2 \mu(dx) < \infty.$$

(iii) The function $\varphi(\cdot, \theta)$ is $L^2(\mu)$-continuous in Θ. That is, for all $\theta \in \Theta$,

$$\int_{\mathbb{R}^n} |\varphi(x, \theta + h) - \varphi(x, \theta)|^2 \mu(dx) \longrightarrow 0 \quad \text{as } h \to 0.$$

Remark 7.4.4 Assumption (ii) can be replaced by the fact that the function $p^{1/2}(x, \theta)$ is $L^2(\mu)$-differentiable in Θ, but the above condition is easier to check in practice.

Example 7.4.5 Assume that the components of F are independent and $N(\theta, 1)$ random variables. This model is regular with, as dominating measure, the Lebesgue measure in $\mathbb{R}^n$.

Example 7.4.6 Consider a Brownian motion with drift on $[0, 1]$,

$$X_t = B_t + \int_0^t b(s, \theta) ds,$$

where, for all $\theta \in \Theta$, $b(\cdot, \theta)$ is a continuously differentiable function in $[0, 1]$. We denote by $(P_\theta)_{\theta \in \Theta}$ the law of the process $X = (X_t)_{t \in [0,1]}$ on the canonical space $(C([0, 1]), \mathcal{B}([0, 1]))$, and we assume that we have observed the process X_t during the whole time interval $[0, 1]$. The definition of a parametric statistical regular model can be extended to this family of probability measures. In this case, the dominating measure μ is the law of Brownian motion on $(C([0, 1]), \mathcal{B}([0, 1]))$. In fact, by Girsanov's theorem (Theorem 2.10.3),

$$\frac{dP_\theta}{d\mu}(X) = \exp\left(\int_0^1 b(t, \theta) dX_t - \frac{1}{2} \int_0^1 b^2(t, \theta) dt \right) =: p(X, \theta).$$

Moreover, this model is regular with $L^2(\mu)$-continuous derivative given by

$$\varphi(X, \theta) = p^{1/2}(X, \theta)\left(\frac{1}{2}\int_0^1 \partial_\theta b(t, \theta)dX_t - \frac{1}{2}\int_0^1 b(t, \theta)\partial_\theta b(t, \theta)dt\right).$$

The functions $p(x, \theta)$ and $\log p(x, \theta)$ are called, respectively, the likelihood and log-likelihood functions. The following is a consequence of the regularity condition. See Ibragimov and Has'minskii (1981, Lemma 7.2) for a proof of this result.

Lemma 7.4.7 *Let $(F_\theta)_{\theta\in\Theta}$ be a regular parametric statistical model. Let $f: \mathbb{R}^n \to \Theta$ be a measurable function such that $E(|f(F_\theta)|^2)$ is bounded in Θ. Then the function $E(f(F_\theta))$ is continuously differentiable in Θ and*

$$\partial_\theta E(f(F_\theta)) = \partial_\theta \int_{\mathbb{R}^n} f(x)p(x, \theta)\mu(dx) = \int_{\mathbb{R}^n} f(x)\partial_\theta p(x, \theta)\mu(dx).$$

We now assume that $(\Omega, \mathcal{F}, P)$ is a probability space, where there exists a Brownian motion B such that $\mathcal{F}$ is generated by B. We next derive a formula for the derivative of the log-likelihood density $\log p(x, \theta)$ that is analogous to (7.16).

Proposition 7.4.8 *Let $(F_\theta)_{\theta\in\Theta}$ be a regular parametric statistical model, where the components of the the n-dimensional random vector F_θ belong to $\mathbb{D}^{1,2}$. Let $(u_t)_{t\geq 0}$ be a process in Dom δ such that, for all $j \in \{1, \ldots, n\}$,*

$$\langle DF_{\theta,j}, u\rangle_H = \partial_\theta F_{\theta,j}. \tag{7.18}$$

Moreover, assume that $E\left(|\partial_\theta F_{\theta,j}|^p\right)$ is uniformly bounded on Θ for some $p > 1$. Then

$$\partial_\theta \log p(x, \theta) = E(\delta(u)|F_\theta = x),$$

for P^θ-almost-all x.

Proof Let $\varphi \in C_0^\infty(\mathbb{R}^n)$. By the chain rule (Proposition 3.3.2) and the duality relationship (Proposition 3.2.3), we obtain

$$\partial_\theta E(\varphi(F_\theta)) = \sum_{j=1}^n E\left(\partial_j\varphi(F_\theta)\partial_\theta F_{\theta,j}\right) = \sum_{j=1}^n E\left(\partial_j\varphi(F_\theta)\langle DF_{\theta,j}, u\rangle_H\right)$$

$$= E(\langle D(\varphi(F_\theta)), u\rangle_H) = E(\varphi(F_\theta)\delta(u)). \tag{7.19}$$

Observe that the fact that $E\left(|\partial_\theta F_{\theta,j}|^p\right)$ is uniformly bounded on Θ implies the uniform integrability of the family $(\partial_\theta F_{\theta,j})_{\theta\in\Theta}$. Therefore, we can interchange the derivative and the expectation in (7.19).

Furthermore, by Lemma 7.4.7,

$$\partial_\theta E(\varphi(F_\theta)) = \int_{\mathbb{R}^n} \varphi(x)\partial_\theta p(x,\theta)\mu(dx)$$

$$= \int_{\mathbb{R}^n} \varphi(x)\partial_\theta \log p(x,\theta)p(x,\theta)\mu(dx)$$

$$= E(\varphi(F_\theta)\,\partial_\theta \log p(F_\theta,\theta)),$$

which concludes the proof. $\qquad\qquad\square$

Example 7.4.9 Assume that $F_{\theta,j} = \epsilon_j + \theta$ for $j = 1,\ldots,n$, where the ϵ_j are independent $N(0,1)$ random variables. Consider orthonormal elements $e_j \in H = L^2(\mathbb{R}_+)$, $j = 1,\ldots,n$. We can assume that $B(e_j) = \epsilon_j$ for $j = 1,\ldots,n$. Then $\langle e_j,e_k\rangle_H = E(\epsilon_j\epsilon_k) = \delta_{jk}$, and $DF_{\theta,j} = e_j$. Let $u = \sum_{j=1}^n e_j$; u clearly satisfies (7.18) since $\partial_\theta F_{\theta,j} = 1$. Moreover,

$$\delta(u) = \sum_{j=1}^n B(e_j) = \sum_{j=1}^n \epsilon_j = \sum_{j=1}^n F_{\theta,j} - n\theta.$$

By Proposition 7.4.8, we obtain $\partial_\theta \log p(x,\theta) = \sum_{j=1}^n x_j - n\theta$.

Example 7.4.10 Let $X^{(n)} = (X_{t_1},\ldots,X_{t_n})$, $0 < t_1 < \cdots < t_n < T$, be observations of the Ornstein–Uhlenbeck process

$$dX_t = -\theta X_t dt + dB_t, \quad X_0 = 0,$$

where B is a Brownian motion on $[0,T]$. To simplify, we have omitted the dependence of $X^{(n)}$ on θ. We have

$$X_t = \int_0^t e^{-\theta(t-s)}dB_s.$$

Therefore $D_s X_t = e^{-\theta(t-s)}\mathbf{1}_{[0,t]}(s)$. Moreover, the process $\partial_\theta X_t$ satisfies

$$d\,\partial_\theta X_t = -X_t dt - \theta\partial_\theta X_t dt, \quad \partial_\theta X_0 = 0.$$

Thus,

$$\partial_\theta X_t = -\int_0^t e^{-\theta(t-s)}X_s ds = -\int_0^T X_s D_s X_t ds = -\langle X, DX_t\rangle_{L^2([0,T])}.$$

Then, by Proposition 7.4.8, taking $u = X$ we obtain

$$\partial_\theta \log p(x,\theta) = -E(\delta(X)|X^{(n)} = x) = -E\left(\int_0^T X_s dB_s \middle| X^{(n)} = x\right)$$

$$= -E\left(\int_0^T X_s dX_s + \theta\int_0^T X_s^2 ds \middle| X^{(n)} = x\right).$$

In particular, the maximum likelihood estimator of θ is given by

$$\hat{\theta} = -\frac{E\left(\int_0^T X_s dX_s \big| X^{(n)}\right)}{E\left(\int_0^T X_s^2 ds \big| X^{(n)}\right)}.$$

These formulas are used in Gobet (2001, 2002) to derive asymptotic results for this parametric model and more general ones.

7.5 Malliavin Differentiability of Diffusion Processes

Suppose that $B = (B_t)_{t \geq 0}$, with $B_t = (B_t^1, \dots, B_t^d)$, is a d-dimensional Brownian motion. Consider the m-dimensional stochastic differential equation

$$dX_t = \sum_{j=1}^d \sigma_j(X_t) dB_t^j + b(X_t) dt, \tag{7.20}$$

with initial condition $X_0 = x_0 \in \mathbb{R}^m$, where the coefficients $\sigma_j, b \colon \mathbb{R}^m \to \mathbb{R}^m$, $1 \leq j \leq d$ are measurable functions.

By definition, a solution to equation (7.20) is an adapted process $X = (X_t)_{t \geq 0}$ such that, for any $T > 0$ and $p \geq 2$,

$$E\left(\sup_{t \in [0,T]} |X_t|^p\right) < \infty$$

and X satisfies the integral equation

$$X_t = x_0 + \sum_{j=1}^d \int_0^t \sigma_j(X_s) dB_s^j + \int_0^t b(X_s) ds. \tag{7.21}$$

The following result is well known (see, for instance, Karatzas and Shreve, 1998).

Theorem 7.5.1 *Suppose that the coefficients $\sigma_j, b \colon \mathbb{R}^m \to \mathbb{R}^m$, $1 \leq j \leq d$, satisfy the Lipschitz condition: for all $x, y \in \mathbb{R}^m$,*

$$\max_j \left(|\sigma_j(x) - \sigma_j(y)|, |b(x) - b(y)|\right) \leq K|x - y|. \tag{7.22}$$

Then there exists a unique solution X to Equation (7.21).

When the coefficients in equation (7.20) are continuously differentiable, the components of the solution are differentiable in the Malliavin calculus sense.

Proposition 7.5.2 *Suppose that the coefficients σ_j, b are in $C^1(\mathbb{R}^m; \mathbb{R}^m)$ and have bounded partial derivatives. Then, for all $t \geq 0$ and $i = 1, \ldots, m$, $X_t^i \in \mathbb{D}^{1,\infty}$, and for $r \leq t$ and $j = 1, \ldots, d$,*

$$D_r^j X_t = \sigma_j(X_r) + \sum_{k=1}^{m} \sum_{\ell=1}^{d} \int_r^t \partial_k \sigma_\ell(X_s) D_r^j X_s^k dB_s^\ell$$

$$+ \sum_{k=1}^{m} \int_r^t \partial_k b(X_s) D_r^j X_s^k ds. \tag{7.23}$$

Proof To simplify, we assume that $b = 0$. Consider the Picard approximations given by $X_t^{(0)} = x_0$ and

$$X_t^{(n+1)} = x_0 + \sum_{j=1}^{d} \int_0^t \sigma_j(X_s^{(n)}) dB_s^j,$$

if $n \geq 0$. We will prove the following claim by induction on n:

Claim: $X_t^{(n),i} \in \mathbb{D}^{1,\infty}$ for all $i = 1, \ldots, m, t \geq 0$. Moreover, for all $p > 1$ and $t \geq 0$,

$$\psi_n(t) := \sup_{0 \leq r \leq t} E\left(\sup_{s \in [r,t]} |D_r X_s^{(n)}|^p \right) < \infty \tag{7.24}$$

and, for all $T > 0$ and $t \in [0, T]$,

$$\psi_{n+1}(t) \leq c_1 + c_2 \int_0^t \psi_n(s) ds, \tag{7.25}$$

for some constants c_1, c_2 depending on T.

Clearly, the claim holds for $n = 0$. Suppose that it is true for n. Applying property (3.6) of the divergence operator and the chain rule (Proposition 3.3.2), for any $r \leq t, i = 1, \ldots, m$, and $\ell = 1, \ldots, d$, we get

$$D_r^\ell X_t^{(n+1),i} = D_r^\ell \left(\sum_{j=1}^{m} \int_0^t \sigma_j^i(X_s^{(n)}) dB_s^j \right)$$

$$= \sum_{j=1}^{m} \left(\delta_{\ell,j} \sigma_\ell^i(X_r^{(n)}) + \int_r^t D_r^\ell \left(\sigma_j^i(X_s^{(n)}) \right) dB_s^j \right)$$

$$= \sum_{j=1}^{m} \left(\delta_{\ell,j} \sigma_\ell^i(X_r^{(n)}) + \sum_{k=1}^{m} \int_r^t \partial_k \sigma_j(X_s^{(n)}) D_r^\ell X_s^{(n),k} dB_s^j \right).$$

From these equalities and condition (7.24) we see that $X_t^{(n+1),i} \in \mathbb{D}^{1,\infty}$ and we obtain, using the Burkholder–David–Gundy inequality (Theorem 2.2.3) and Hölder's inequality,

$$E\left(\sup_{r \le s \le t} |D_r X_s^{(n+1)}|^p\right) \le c_p \left(\gamma_p + T^{(p-1)/2} K^p \int_r^t E(|D_r^j X_s^{(n)}|^p) ds\right), \quad (7.26)$$

where

$$\gamma_p = \sup_{n,j} E\left(\sup_{0 \le t \le T} |\sigma_j(X_t^{(n)})|^p\right) < \infty.$$

So (7.24) and (7.25) hold for $n + 1$ and the claim is proved.

We know that

$$E\left(\sup_{s \le T} |X_s^{(n)} - X_s|^p\right) \longrightarrow 0$$

as n tends to infinity. By Gronwall's lemma applied to (7.25) we deduce that the derivatives of the sequence $X_t^{(n),i}$ are bounded in $L^p(\Omega; H)$ uniformly in n for all $p \ge 2$. This implies that the random variables X_t^i belong to $\mathbb{D}^{1,\infty}$. Finally, applying the operator D to equation (7.21) we deduce the linear stochastic differential equation (7.23) for the derivative of X_t^i.

This completes the proof of the proposition.   □

Example 7.5.3   Consider the diffusion process in $\mathbb{R}$

$$dX_t = \sigma(X_t)dB_t + b(X_t)dt, \qquad X_0 = x_0,$$

where σ and b are globally Lipschitz functions in $C^1(\mathbb{R})$. Then, for all $t > 0$, X_t belongs to $\mathbb{D}^{1,\infty}$ and the Malliavin derivative $(D_r X_t)_{r \le t}$ satisfies the following linear equation:

$$D_r X_t = \sigma(X_r) + \int_r^t \sigma'(X_s)D_r(X_s)dB_s + \int_r^t b'(X_s)D_r(X_s)ds.$$

Therefore, by Itô's formula (Theorem 2.4.3),

$$D_r X_t = \sigma(X_t) \exp\left(\int_r^t \sigma'(X_s)dB_s + \int_r^t (b(X_s) - \tfrac{1}{2}(\sigma')^2(X_s))ds\right).$$

Consider the $m \times m$ matrix-valued process defined by

$$Y_t = I_m + \sum_{l=1}^d \int_0^t \partial\sigma_\ell(X_s)Y_s dB_s^\ell + \int_0^t \partial b(X_s)Y_s ds,$$

where I_m denotes the identity matrix of order m and $\partial\sigma_\ell$ denotes the $m \times m$ Jacobian matrix of the function σ_ℓ; that is,

$$(\partial\sigma_\ell)_j^i = \partial_j \sigma_\ell^i.$$

In the same way, ∂b denotes the $m \times m$ Jacobian matrix of b. If the coefficients of equation (7.21) are of class $C^{1+\alpha}$, $\alpha > 0$, then there is a version of the solution $X_t(x_0)$ to this equation that is continuously differentiable in x_0 (see Kunita, 1984), and for which Y_t is the Jacobian matrix $\partial X_t / \partial x_0$:

$$Y_t = \frac{\partial X_t}{\partial x_0}.$$

Proposition 7.5.4 *For any $t \in [0, T]$ the matrix Y_t is invertible. Its inverse Z_t satisfies*

$$Z_t = I_m - \sum_{\ell=1}^{d} \int_0^t Z_s \partial\sigma_\ell(X_s) dB_s^\ell$$

$$- \int_0^t Z_s \left(\partial b(X_s) - \sum_{\ell=1}^{d} \partial\sigma_\ell(X_s)\partial\sigma_\ell(X_s) \right) ds.$$

Proof By means of Itô's formula (Theorem 2.4.3), one can check that $Z_t Y_t = Y_t Z_t = I_m$, which implies that $Z_t = Y_t^{-1}$. In fact,

$$Z_t Y_t = I_m + \sum_{\ell=1}^{d} \int_0^t Z_s \, \partial\sigma_\ell(X_s) Y_s dB_s^\ell + \int_0^t Z_s \partial b(X_s) Y_s ds$$

$$- \sum_{\ell=1}^{d} \int_0^t Z_s \partial\sigma_l(X_s) Y_s dB_s^\ell$$

$$- \int_0^t Z_s \left(\partial b(X_s) - \sum_{\ell=1}^{d} \partial\sigma_\ell(X_s)\partial\sigma_\ell(X_s) \right) Y_s ds$$

$$- \int_0^t Z_s \left(\sum_{\ell=1}^{d} \partial\sigma_\ell(X_s)\partial\sigma_\ell(X_s) \right) Y_s ds = I_m.$$

Similarly, we can show that $Y_t Z_t = I_m$. □

Lemma 7.5.5 *The $m \times d$ matrix $(D_r X_t)_j^i = D_r^j X_t^i$ can be expressed as*

$$D_r X_t = Y_t Y_r^{-1} \sigma(X_r), \tag{7.27}$$

where σ denotes the $m \times d$ matrix with columns $\sigma_1, \ldots, \sigma_d$.

Proof It suffices to check that the process $\Phi_{t,r} := Y_t Y_r^{-1} \sigma(X_r)$, $t \geq r$ satisfies

$$\Phi_{t,r} = \sigma(X_r) + \sum_{\ell=1}^{d} \int_r^t \partial\sigma_\ell(X_s)\Phi_{s,r} dB_s^\ell + \int_r^t \partial b(X_s)\Phi_{s,r} ds.$$

In fact,

$$\sigma(X_r) + \sum_{\ell=1}^{d} \int_r^t \partial \sigma_\ell(X_s)(Y_s Y_r^{-1}\sigma(X_r))dB_s^\ell$$

$$+ \int_r^t \partial b(X_s)(Y_s Y_r^{-1}\sigma(X_r))ds$$

$$= \sigma(X_r) + (Y_t - Y_r)Y_r^{-1}\sigma(X_r) = Y_t Y_r^{-1}\sigma(X_r).$$

This completes the proof. □

Consider the Malliavin matrix of X_t, denoted by $\gamma_{X_t} := Q_t$ and given by

$$Q_t^{i,j} = \sum_{\ell=1}^{d} \int_0^t D_s^\ell X_t^i D_s^\ell X_t^j ds.$$

That is, $Q_t = \int_0^t (D_s X_t)(D_s X_t)^T ds$. Equation (7.27) leads to

$$Q_t = Y_t C_t Y_t^T, \tag{7.28}$$

where

$$C_t = \int_0^t Y_s^{-1}\sigma\sigma^T(X_s)(Y_s^{-1})^T ds.$$

Taking into account that Y_t is invertible, the nondegeneracy of the matrix Q_t will depend only on the nondegeneracy of the matrix C_t, which is called the *reduced Malliavin matrix*.

7.6 Absolute Continuity under Ellipticity Conditions

Consider the stopping time defined by

$$S = \inf\{t > 0 : \det \sigma\sigma^T(X_t) \neq 0\}.$$

Theorem 7.6.1 (Bouleau and Hirsch, 1986) *Let $(X_t)_{t \geq 0}$ be a diffusion process with $C^{1+\alpha}$ and Lipschitz coefficients. Then, for any $t > 0$, the law of X_t conditioned by $\{t > S\}$ is absolutely continuous with respect to the Lebesgue measure on $\mathbb{R}^m$.*

Proof It suffices to show that $\det C_t > 0$ a.s. on the set $\{S < t\}$. Suppose that $t > S$. For any $u \in \mathbb{R}^m$ with $|u| = 1$ we can write

$$u^T C_t u = \int_0^t u^T Y_s^{-1}\sigma\sigma^T(X_s)(Y_s^{-1})^T u ds$$

$$\geq \int_0^t \inf_{|v|=1} \left(v^T \sigma\sigma^T(X_s)v \right) |(Y_s^{-1})^T u|^2 ds.$$

Notice that $\inf_{|v|=1}\left(v^T\sigma\sigma^T(X_s)v\right)$ is the smallest eigenvalue of $\sigma\sigma^T(X_s)$, which is strictly positive in an open interval contained in $[0,t]$ by the definition of the stopping time S and because $t > S$.

Furthermore, $|(Y_s^{-1})^T u| \geq |u|\,|Y_s|^{-1}$. Therefore we obtain

$$u^T C_t u \geq k|u|^2,$$

for some positive random variable $k > 0$, which implies that the matrix C_t is invertible. This completes the proof. $\qquad\square$

Example 7.6.2 Assume that $\sigma(x_0) \neq 0$ in Example 7.5.3. Then, for any $t > 0$, the law of X_t is absolutely continuous with respect to the Lebesgue measure in $\mathbb{R}$.

7.7 Regularity of the Density under Hörmander's Conditions

We need the following regularity result, whose proof is similar to that of Proposition 7.5.2 and is thus omitted.

Proposition 7.7.1 *Suppose that the coefficients σ_j, $1 \leq j \leq m$, and b of equation (7.20) are infinitely differentiable with bounded derivatives of all orders. Then, for all $t \geq 0$ and $i = 1,\ldots,m$, X_t^i belong to $\mathbb{D}^\infty$.*

Consider the following vector fields on $\mathbb{R}^m$:

$$\sigma_j = \sum_{i=1}^m \sigma_j^i(x)\frac{\partial}{\partial x_i}, \quad j = 1,\ldots,d,$$

$$b = \sum_{i=1}^m b^i(x)\frac{\partial}{\partial x_i}.$$

The Lie bracket between the vector fields σ_j and σ_k is defined by

$$[\sigma_j,\sigma_k] = \sigma_j\sigma_k - \sigma_k\sigma_j = \sigma_j^\nabla\sigma_k - \sigma_k^\nabla\sigma_j,$$

where

$$\sigma_j^\nabla\sigma_k = \sum_{i,\ell=1}^m \sigma_j^\ell\partial_\ell\sigma_k^i\frac{\partial}{\partial x_i}.$$

Set

$$\sigma_0 = b - \frac{1}{2}\sum_{\ell=1}^d \sigma_\ell^\nabla\sigma_\ell.$$

The vector field σ_0 appears when we write the stochastic differential equation (7.21) in terms of the Stratonovich integral (see Section 2.7) instead of Itô's integral:

$$X_t = x_0 + \sum_{j=1}^{d} \int_0^t \sigma_j(X_s) \circ dB_s^j + \int_0^t \sigma_0(X_s) ds.$$

Let us introduce the nondegeneracy condition required for the smoothness of the density.

(**HC**) *Hörmander's condition:* The vector space spanned by the vector fields

$$\sigma_1, \ldots, \sigma_d, [\sigma_i, \sigma_j], 0 \leq i \leq d, 1 \leq j \leq d, [\sigma_i, [\sigma_j, \sigma_k]], 0 \leq i, j, k \leq d, \ldots$$

at the point x_0 is $\mathbb{R}^m$.

For instance, if $m = d = 1$, $\sigma_1^1(x) = a(x)$ and $\sigma_0^1(x) = a_0(x)$; then Hörmander's condition means that $a(x_0) \neq 0$ or $a^n(x_0)a_0(x_0) \neq 0$ for some $n \geq 1$.

Theorem 7.7.2 *Assume that Hörmander's condition holds. Then, for any $t > 0$, the random vector X_t has an infinitely differentiable density.*

This result can be considered as a probabilistic version of Hörmander's theorem on the hypoellipticity of second-order differential operators. In fact, the density p_t of X_t satisfies the Fokker–Planck equation

$$\left(-\frac{\partial}{\partial t} + L^*\right) p_t = 0,$$

where

$$L = \frac{1}{2} \sum_{i,j=1}^{m} (\sigma \sigma^T)^{ij} \frac{\partial^2}{\partial x_i \partial x_j} + \sum_{i=1}^{m} b^i \frac{\partial}{\partial x_i}.$$

Then, $p_t \in C^\infty(\mathbb{R}^m)$ means that $\partial/\partial t - L^*$ is hypoelliptic (Hörmander's theorem).

For the proof of Theorem 7.7.2 we need several technical lemmas.

Lemma 7.7.3 *Let C be an $m \times m$ symmetric nonnegative definite random matrix. Assume that the entries C^{ij} have moments of all orders and that for any $p \geq 2$ there exists $\epsilon_0(p)$ such that, for all $\epsilon \leq \epsilon_0(p)$,*

$$\sup_{|v|=1} P\left(v^T C v \leq \epsilon\right) \leq \epsilon^p.$$

Then $E((\det C)^{-p}) < \infty$ for all $p \geq 2$.

Proof Let $\lambda = \inf_{|v|=1} v^T C v$ be the smallest eigenvalue of C. We know that $\lambda^m \leq \det C$. Thus, it suffices to show that $E(\lambda^{-p}) < \infty$ for all $p \geq 2$. Set $|C| = \left(\sum_{i,j=1}^m (C^{ij})^2\right)^{\frac{1}{2}}$. Fix $\epsilon > 0$, and let $v_1, \ldots, v_N$ be a finite set of unit vectors such that the balls with their center in these points and radius $\epsilon^2/2$ cover the unit sphere S^{m-1}. Then, we have

$$P(\lambda < \epsilon) = P\left(\inf_{|v|=1} v^T C v < \epsilon\right)$$

$$\leq P\left(\inf_{|v|=1} v^T C v < \epsilon, |C| \leq \frac{1}{\epsilon}\right) + P\left(|C| > \frac{1}{\epsilon}\right). \quad (7.29)$$

Assume that $|C| \leq 1/\epsilon$ and $v_k^T C v_k \geq 2\epsilon$ for any $k = 1, \ldots, N$. For any unit vector v, there exists a v_k such that $|v - v_k| \leq \epsilon^2/2$ and we can deduce the following inequalities:

$$v^T C v \geq v_k^T C v_k - |v^T C v - v_k^T C v_k|$$
$$\geq 2\epsilon - (|v^T C v - v^T C v_k| + |v^T C v_k - v_k^T C v_k|)$$
$$\geq 2\epsilon - 2|C| |v - v_k| \geq \epsilon.$$

As a consequence, (7.29) implies that

$$P(\lambda < \epsilon) \leq P\left(\bigcup_{k=1}^N \{v_k^T C v_k < 2\epsilon\}\right) + P\left(|C| > \frac{1}{\epsilon}\right) \leq N(2\epsilon)^{p+2m} + \epsilon^p E(|C|^p)$$

if $\epsilon \leq \frac{1}{2}\epsilon_0(p + 2m)$. The number N depends on ϵ but is bounded by a constant times ϵ^{-2m}. Therefore, we obtain $P(\lambda < \epsilon) \leq C\epsilon^p$ for all $\epsilon \leq \epsilon_1(p)$ and for all $p \geq 2$. This implies that λ^{-1} has moments of all orders, which completes the proof of the lemma. $\quad \square$

Lemma 7.7.4 *Let $(Z_t)_{t\geq 0}$ be a real-valued, adapted, continuous process such that $Z_0 = z_0 \neq 0$. Suppose that there exist $\alpha > 0$ and $t_0 > 0$ such that, for all $p \geq 1$ and $t \in [0, t_0]$,*

$$E\left(\sup_{0\leq s\leq t} |Z_s - z_0|^p\right) \leq C_p t^{p\alpha}.$$

Then, for all $p \geq 1$ and $t \geq 0$,

$$E\left(\left(\int_0^t |Z_s| ds\right)^{-p}\right) < \infty.$$

Proof   We can assume that $t \in [0, t_0]$. For any $0 < \epsilon < t|z_0|/2$, we have

$$P\left(\int_0^t |Z_s|ds < \epsilon\right) \le P\left(\int_0^{2\epsilon/|z_0|} |Z_s|ds < \epsilon\right)$$

$$\le P\left(\sup_{0 \le s \le 2\epsilon/|z_0|} |Z_s - z_0| > \frac{|z_0|}{2}\right)$$

$$\le \frac{2^p C_p}{|z_0|^p}\left(\frac{2\epsilon}{|z_0|}\right)^{p\alpha},$$

which implies the desired result.   □

The next lemma was proved by Norris (1986), following the ideas of Stroock (1983), and is the basic ingredient in the proof of Theorem 7.7.2.

Lemma 7.7.5 (Norris's lemma)   *Consider a continuous semimartingale of the form*

$$Y_t = y + \int_0^t a_s ds + \sum_{i=1}^d \int_0^t u_s^i dB_s^i,$$

where

$$a(t) = \alpha + \int_0^t \beta_s ds + \sum_{i=1}^d \int_0^t \gamma_s^i dB_s^i$$

and $c = E\left(\sup_{0 \le t \le T}(|\beta_t| + |\gamma_t| + |a_t| + |u_t|)^p\right) < \infty$ for some $p \ge 2$.
   Fix $q > 8$. Then, for all $r < (q - 8)/27$ there exists an ϵ_0 such that, for all $\epsilon \le \epsilon_0$, we have

$$P\left(\int_0^T Y_t^2 dt < \epsilon^q, \int_0^T (|a_t|^2 + |u_t|^2)dt \ge \epsilon\right) \le c_1 \epsilon^{rp}.$$

Proof of Theorem 7.7.2   The proof will be carried out in several steps:

Step 1   We need to show that, for all $t > 0$ and all $p \ge 2$, $E((\det Q_t)^{-p}) < \infty$, where Q_t is the Malliavin matrix of X_t. Taking into account that

$$E\left(|\det Y_t^{-1}|^p + |\det Y_t|^p\right) < \infty,$$

it suffices to show that $E((\det C_t)^{-p}) < \infty$ for all $p \ge 2$.

Step 2   Fix $t > 0$. Using Lemma 7.7.3, the problem reduces to showing that, for all $p \ge 2$, we have

$$\sup_{|v|=1} P\left(v^T C_t v \le \epsilon\right) \le \epsilon^p,$$

for any $\epsilon \leq \epsilon_0(p)$, where the quadratic form associated with the matrix C_t is given by

$$v^T C_t v = \sum_{j=1}^{d} \int_0^t \langle v, Y_s^{-1} \sigma_j(X_s) \rangle^2 ds. \tag{7.30}$$

Step 3 Fix a smooth function V and use Itô's formula to compute the differential of $Y_t^{-1} V(X_t)$:

$$d\left(Y_t^{-1} V(X_t)\right) = Y_t^{-1} \sum_{k=1}^{d} [\sigma_k, V](X_t) dB_t^k$$

$$+ Y_t^{-1}\left([\sigma_0, V] + \frac{1}{2} \sum_{k=1}^{d} [\sigma_k, [\sigma_k, V]]\right)(X_t) dt. \tag{7.31}$$

Step 4 We introduce the following sets of vector fields:

$$\Sigma_0 = \{\sigma_1, \ldots, \sigma_d\},$$
$$\Sigma_n = \{[\sigma_k, V], k = 0, \ldots, d, V \in \Sigma_{n-1}\} \quad \text{if } n \geq 1,$$
$$\Sigma = \cup_{n=0}^{\infty} \Sigma_n$$

and

$$\Sigma_0' = \Sigma_0,$$
$$\Sigma_n' = \left\{[\sigma_k, V], k = 1, \ldots, d, V \in \Sigma_{n-1}';\right.$$

$$\left.[\sigma_0, V] + \frac{1}{2} \sum_{j=1}^{d} [\sigma_j, [\sigma_j, V]], V \in \Sigma_{n-1}'\right\} \quad \text{if } n \geq 1,$$
$$\Sigma' = \cup_{n=0}^{\infty} \Sigma_n'.$$

We denote by $\Sigma_n(x)$ (resp. $\Sigma_n'(x)$) the subset of $\mathbb{R}^m$ obtained by freezing the variable x in the vector fields of Σ_n (resp. Σ_n'). Clearly, the vector spaces spanned by $\Sigma(x_0)$ or by $\Sigma'(x_0)$ coincide and, under Hörmander's condition, this vector space is $\mathbb{R}^m$. Therefore, there exists an integer $j_0 \geq 0$ such that the linear span of the set of vector fields $\cup_{j=0}^{j_0} \Sigma_j'(x)$ at point x_0 has dimension m.

As a consequence there exist constants $R > 0$ and $c > 0$ such that

$$\sum_{j=0}^{j_0} \sum_{V \in \Sigma_j'} \langle v, V(y) \rangle^2 \geq c, \tag{7.32}$$

for all v and y with $|v| = 1$ and $|y - x_0| < R$.

Step 5   For any $j = 0, 1, \ldots, j_0$ we put $m(j) = 2^{-4j}$ and define the set

$$E_j = \left\{ \sum_{V \in \Sigma'_j} \int_0^t \langle v, Y_s^{-1} V(X_s) \rangle^2 ds \le \epsilon^{m(j)} \right\}.$$

Notice that $\{v^T C_t v \le \epsilon\} = E_0$ because $m(0) = 1$. Consider the decomposition

$$E_0 \subset (E_0 \cap E_1^c) \cup (E_1 \cap E_2^c) \cup \cdots \cup (E_{j_0-1} \cap E_{j_0}^c) \cup F,$$

where $F = E_0 \cap E_1 \cap \cdots \cap E_{j_0}$. Then, for any unit vector v, we have

$$P(v^T C_t v \le \epsilon) = P(E_0) \le P(F) + \sum_{j=0}^{j_0-1} P(E_j \cap E_{j+1}^c).$$

We will now estimate each term in this sum.

Step 6   Let us first estimate $P(F)$. By the definition of F we obtain

$$P(F) \le P\left(\sum_{j=0}^{j_0} \sum_{V \in \Sigma'_j} \int_0^t \langle v, Y_s^{-1} V(X_s) \rangle^2 ds \le (j_0 + 1)\epsilon^{m(j_0)} \right).$$

Then, taking into account (7.32), we can apply Lemma 7.7.4 to the process

$$Z_s = \inf_{|v|=1} \sum_{j=0}^{j_0} \sum_{V \in \Sigma'_j} \langle v, Y_s^{-1} V(X_s) \rangle^2,$$

and we obtain

$$E\left(\left| \inf_{|v|=1} \sum_{j=0}^{j_0} \sum_{V \in \Sigma'_j} \int_0^t \langle v, Y_s^{-1} V(X_s) \rangle^2 ds \right|^{-p} \right) < \infty.$$

Therefore, for any $p \ge 1$, there exists ϵ_0 such that

$$P(F) \le \epsilon^p$$

for any $\epsilon < \epsilon_0$.

Step 7   For any $j = 0, \ldots, j_0$, the probability of the event $E_j \cap E_{j+1}^c$ is bounded by the sum with respect to $V \in \Sigma'_j$ of the probability that the two following events happen:

$$\int_0^t \langle v, Y_s^{-1} V(X_s) \rangle^2 ds \le \epsilon^{m(j)}$$

and

$$\sum_{k=1}^{d} \int_0^t \langle v, Y_s^{-1}[\sigma_k, V](X_s) \rangle^2 ds$$

$$+ \int_0^t \left\langle v, Y_s^{-1}\left([\sigma_0, V] + \frac{1}{2}\sum_{j=1}^{d}[\sigma_j, [\sigma_j, V]]\right)(X_s)\right\rangle^2 ds > \frac{\epsilon^{m(j+1)}}{n(j)},$$

where $n(j)$ denotes the cardinality of the set Σ'_j.

Consider the continuous semimartingale $(\langle v, Y_s^{-1}V(X_s)\rangle)_{s\geq0}$. From (7.31) we see that the quadratic variation of this semimartingale is equal to

$$\sum_{k=1}^{d} \int_0^s \langle v, Y_r^{-1}[\sigma_k, V](X_r) \rangle^2 dr,$$

and the bounded variation component is

$$\int_0^s \left\langle v, Y_r^{-1}\left([\sigma_0, V] + \frac{1}{2}\sum_{j=1}^{d}[\sigma_j, [\sigma_j, V]]\right)(X_r)\right\rangle dr.$$

Taking into account that $8m(j + 1) < m(j)$, from Norris's lemma (Lemma 7.7.5) applied to the semimartingale $Y_s = v^T Y_s^{-1}V(X_s)$, we get that, for any $p \geq 1$, there exists an $\epsilon_0 > 0$ such that

$$P(E_j \cap E_{j+1}^c) \leq \epsilon^p,$$

for all $\epsilon \leq \epsilon_0$. The proof of the theorem is now complete. $\square$

Exercises

7.1 Let F be a random variable with the law $N(0, \sigma^2)$. Show that the density $p(x)$ of F satisfies

$$p(x) = \sigma^{-2}E\left(1_{\{F>x\}}F\right).$$

7.2 Derive (7.4) applying (7.3) and Meyer's inequality (5.7).

7.3 Set $M_t = \int_0^t u_s dB_s$, where $u = (u_t)_{t\in(0,T]}$ is a process in $L_T^2(\mathcal{P})$ such that, for all $t \in [0, T]$, $|u_t| \geq \rho > 0$ for some constant ρ, $u_t \in \mathbb{D}^{2,p}$, and

$$\lambda := \sup_{s,t\in[0,T]} E(|D_s u_t|^p) + \sup_{r,s\in[0,T]} E\left(\left(\int_0^T |D_{r,s}^2 u_t|^2 dt\right)^{p/2}\right) < \infty,$$

for some $p > 3$. Applying Proposition 7.1.6 show that the density of M_t, denoted by $p_t(x)$, satisfies

$$p_t(x) \leq \frac{c}{\sqrt{t}}P(|M_t| > |x|)^{1/q},$$

for all $t \in [0, T]$, where $q > p/(p - 3)$ and the constant c depends on λ, ρ, and p.

Hint: Use a lower bound of the form

$$\|DM_t\|_H^2 = \int_0^t \left(u_s + \int_s^t D_s u_r dB_r \right)^2 ds \geq \int_{t(1-h)}^t \left(u_s + \int_s^t D_s u_r dB_r \right)^2 ds$$

and choose h small enough.

7.4   Show inequality (7.8).

7.5   Consider the parametric model $F_{\theta,j} = \theta \epsilon_j$, $j = 1, \ldots, n$, where the ϵ_j are independent $N(0, 1)$ random variables. With the same notation as in Example 7.4.5 and using Proposition 7.4.8, derive an expression for $\partial_\theta \log p(x, \theta)$.

7.6   Consider the stochastic differential equation

$$dX_t^1 = dB_t^1 + \sin X_t^2 dB_t^2,$$
$$dX_t^2 = 2X_t^1 dB_t^1 + X_t^1 dB_t^2,$$

with initial condition $x_0 = 0$. Show that the vector fields σ_1 and $[\sigma_1, \sigma_2]$ at $x = 0$ span $\mathbb{R}^2$ and that Hörmander's condition holds. Deduce that X_t has a $C^\infty(\mathbb{R}^2)$ density for any $t > 0$.

8

Normal Approximations

In this chapter we present the application of Malliavin calculus, combined with Stein's method (see Chen *et al.*, 2011), to normal approximations. We refer the reader to the book by Nourdin and Peccati (2012) for a detailed account of this topic.

8.1 Stein's Method

The following lemma is a characterization of the standard normal distribution on the real line.

Lemma 8.1.1 (Stein's lemma) *A random variable X such that* $E(|X|) < \infty$ *has the standard normal distribution* $N(0, 1)$ *if and only if, for any function* $f \in C_b^1(\mathbb{R})$, *we have*

$$E(f'(X) - f(X)X) = 0. \tag{8.1}$$

Proof Suppose first that X has the standard normal distribution $N(0, 1)$. Then, equality (8.1) follows integrating by parts and using that the density $p(x) = (1/\sqrt{2\pi}) \exp(-x^2/2)$ satisfies the differential equation

$$p'(x) = -xp(x).$$

Conversely, let $\varphi(\lambda) = E(e^{i\lambda X})$, $\lambda \in \mathbb{R}$, be the characteristic function of X. Because X is integrable, we know that φ is differentiable and $\varphi'(\lambda) = iE(Xe^{i\lambda X})$. By our assumption, this is equal to $-\lambda\varphi(\lambda)$. Therefore, $\varphi(\lambda) = \exp(-\lambda^2/2)$, which concludes the proof. $\qquad\square$

If the expectation $E(f'(X) - f(X)X)$ is small for functions f in some large set, we might conclude that the distribution of X is close to the normal distribution. This is the main idea of Stein's method for normal approximations and the goal is to quantify this assertion in a proper way. To do this, consider a random variable X with the $N(0, 1)$ distribution and fix a measurable function $h\colon \mathbb{R} \to \mathbb{R}$ such that $E(|h(X)|) < \infty$. Stein's equation associated with h is the linear differential equation

$$f_h'(x) - xf_h(x) = h(x) - E(h(X)), \quad x \in \mathbb{R}. \tag{8.2}$$

Definition 8.1.2 A solution to equation (8.2) is an absolutely continuous function f_h such that there exists a version of the derivative f_h' satisfying (8.2) for every $x \in \mathbb{R}$.

The next result provides the existence of a unique solution to Stein's equation.

Proposition 8.1.3 *The function*

$$f_h(x) = e^{x^2/2} \int_{-\infty}^x (h(y) - E(h(X)))e^{-y^2/2}dy \tag{8.3}$$

is the unique solution of Stein's equation (8.2) satisfying

$$\lim_{x \to \pm\infty} e^{-x^2/2} f_h(x) = 0. \tag{8.4}$$

Proof Equation (8.2) can be written as

$$e^{x^2/2} \frac{d}{dx}\left(e^{-x^2/2} f_h(x)\right) = h(x) - E(h(X)).$$

This implies that any solution to equation (8.2) is of the form

$$f_h(x) = ce^{x^2/2} + e^{x^2/2} \int_{-\infty}^x (h(y) - E(h(X)))e^{-y^2/2}dy,$$

for some $c \in \mathbb{R}$. Taking into account that

$$\lim_{x \to \pm\infty} \int_{-\infty}^x (h(y) - E(h(X)))e^{-y^2/2}dy = 0,$$

the asymptotic condition (8.4) is satisfied if and only if $c = 0$. $\qquad\square$

Notice that, since $\int_{\mathbb{R}}(h(y) - E(h(X)))e^{-y^2/2}dy = 0$, we have

$$\int_{-\infty}^x (h(y) - E(h(X)))e^{-y^2/2}dy = -\int_x^\infty (h(y) - E(h(X)))e^{-y^2/2}dy. \tag{8.5}$$

Let us go back to the problem of approximating the law of a random variable F by the $N(0, 1)$ law. Consider a measurable function $h\colon \mathbb{R} \to \mathbb{R}$ such that $E(|h(X)|) < \infty$ and $E(|h(F)|) < \infty$, where X is a random variable with law $N(0, 1)$. Let f_h be the unique solution to the Stein equation (8.2) given by (8.3). Substituting x by a random variable F in (8.3) and taking the expectation, we obtain

$$E(h(F) - E(h(X))) = E(f_h'(F) - Ff_h(F)). \tag{8.6}$$

We are going to consider measurable functions h belonging to a *separating class* of functions $\mathcal{H}$, which means that if two random variables F and G are such that $h(F), h(G) \in L^1(\Omega)$ and $E(h(F)) = E(h(G))$, for all $h \in \mathcal{H}$, then F and G have the same law.

Definition 8.1.4 Let $\mathcal{H}$ be a separating class and let F and G be two random variables such that $h(F)$ and $h(G)$ are in $L^1(\Omega)$ for any $h \in \mathcal{H}$. Then, we define

$$d_{\mathcal{H}}(F, G) = \sup_{h \in \mathcal{H}}(|E(h(F)) - E(h(G))|).$$

Notice that $d_{\mathcal{H}}(F, G)$ depends only on the laws of the random variables F and G, and we can also write $d_{\mathcal{H}}(F, G) = d_{\mathcal{H}}(P \circ F^{-1}, P \circ G^{-1})$. It can be proved that $d_{\mathcal{H}}$ is a distance in the class of probability measures ν on $\mathbb{R}$ such $h \in L^1(\nu)$ for any $h \in \mathcal{H}$. The following are important examples of separating classes of functions $\mathcal{H}$ and their associated distances.

(1) When $\mathcal{H}$ is the class of functions of the form $h = \mathbf{1}_B$, where B is a Borel set in $\mathbb{R}$, the distance $d_{\mathcal{H}}$ is the *total variation distance*, denoted by

$$d_{\mathrm{TV}}(F, G) = \sup_{B \in \mathcal{B}(\mathbb{R})} |P(F \in B) - P(G \in B)|.$$

(2) When $\mathcal{H}$ is the class of functions of the form $h = \mathbf{1}_{(-\infty, z]}$, for some $z \in \mathbb{R}$, the distance $d_{\mathcal{H}}$ is the *Kolmogorov distance*, denoted by

$$d_{\mathrm{Kol}}(F, G) = \sup_{z \in \mathbb{R}} |P(F \leq z) - P(G \leq z)|.$$

(3) When $\mathcal{H} = \mathrm{Lip}(1)$ is the class of functions h such that

$$\|h\|_{\mathrm{Lip}} := \sup_{x \neq y} \frac{|h(x) - h(y)|}{|x - y|} \leq 1,$$

the distance $d_{\mathcal{H}}$ is the *Wasserstein distance*, denoted by

$$d_{\mathrm{W}}(F, G) = \sup_{h \in \mathrm{Lip}(1)} |E(h(F)) - E(h(G))|.$$

Then, from (8.6), we deduce the following result.

Proposition 8.1.5 *Let F, X be two random variables such that X has the $N(0, 1)$ distribution. If $\mathcal{H}$ is a separating class of functions such that $E(|h(X)|) < \infty$ and $E(|h(F)|) < \infty$ for every $h \in \mathcal{H}$ then*

$$d_{\mathcal{H}}(F, X) \leq \sup_{h \in \mathcal{H}} |E(f_h'(F) - F f_h(F))|. \tag{8.7}$$

In order to control these distances, we need estimates on the supremum norm of f_h and its derivative, when h belongs to one of the above classes. These will be obtained in the following propositions.

Proposition 8.1.6 *Let $h: \mathbb{R} \to [0, 1]$ be a measurable function. Then the solution to Stein's equation f_h given by (8.3) satisfies*

$$\|f_h\|_\infty \leq \sqrt{\frac{\pi}{2}} \quad and \quad \|f'_h\|_\infty \leq 2. \tag{8.8}$$

Proof Taking into account that $|h(x) - \mathrm{E}(h(X))| \leq 1$, where X has law $N(0, 1)$, we obtain

$$|f_h(x)| \leq e^{x^2/2} \int_{|x|}^\infty e^{-y^2/2} dy = \sqrt{\frac{\pi}{2}},$$

because the function $x \to e^{x^2/2} \int_{|x|}^\infty e^{-y^2/2} dy$ attains its maximum at $x = 0$.

To prove the second estimate, observe that, in view of (8.5), we can write

$$f'_h(x) = h(x) - \mathrm{E}(h(X)) + xe^{x^2/2} \int_{-\infty}^x (h(y) - \mathrm{E}(h(X)))e^{-y^2/2} dy$$

$$= h(x) - \mathrm{E}(h(X)) - xe^{x^2/2} \int_x^\infty (h(y) - \mathrm{E}(h(X)))e^{-y^2/2} dy,$$

for every $x \in \mathbb{R}$. Therefore

$$|f'_h(x)| \leq 1 + |x|e^{x^2/2} \int_{|x|}^\infty e^{-y^2/2} dy = 2.$$

This completes the proof. $\square$

For the class $\{\mathbf{1}_{(-\infty,z]}, z \in \mathbb{R}\}$ we can improve the estimate (8.8) slightly, as follows. For any $z \in \mathbb{R}$, set $f_z = f_h$, where $h = \mathbf{1}_{(-\infty,z]}$. In this case,

$$f_z(x) = \begin{cases} \sqrt{2\pi}e^{x^2/2}\Phi(x)(1 - \Phi(z)) & \text{if } x \leq z, \\ \sqrt{2\pi}e^{x^2/2}\Phi(z)(1 - \Phi(x)) & \text{if } x \geq z, \end{cases}$$

where $\Phi(x) = P(X \leq x)$, and X has law $N(0, 1)$. The proof of the following estimate is left as an exercise (see Exercise 8.1).

Proposition 8.1.7 *For any $z \in \mathbb{R}$, $\|f_z\|_\infty \leq \sqrt{2\pi}/4$ and $\|f'_z\|_\infty \leq 1$.*

Finally, the following results will be useful in obtaining an upper bound for the Wasserstein distance.

Proposition 8.1.8 *Let $h \in \mathrm{Lip}(1)$. Then the function f_h given in (8.3) admits the representation*

$$f_h(x) = -\int_0^\infty \frac{e^{-t}}{\sqrt{1 - e^{-2t}}} \mathrm{E}\left(h\left(xe^{-t} + \sqrt{1 - e^{-2t}}X\right)X\right)dt. \qquad (8.9)$$

Moreover f_h is continuously differentiable and $\|f_h'\|_\infty \leq \sqrt{2/\pi}$.

Proof The fact that $f_h \in C^1(\mathbb{R})$ follows immediately from equation (8.3). Denote by $\tilde{f}_h(x)$ the right-hand side of equation (8.9). By the dominated convergence theorem, we have

$$\tilde{f}_h'(x) = -\int_0^\infty \frac{e^{-2t}}{\sqrt{1 - e^{-2t}}} \mathrm{E}\left(h'\left(xe^{-t} + \sqrt{1 - e^{-2t}}X\right)X\right)dt. \qquad (8.10)$$

Taking into account that $|h'(x)| \leq 1$, we obtain the estimate $\|\tilde{f}_h'\|_\infty \leq \sqrt{2/\pi}$. Then, it suffices to show that $\tilde{f}_h = f_h$. Applying Lemma 8.1.1 yields

$$\tilde{f}_h(x) = -\int_0^\infty e^{-t}\mathrm{E}\left(h'\left(xe^{-t} + \sqrt{1 - e^{-2t}}X\right)\right)dt. \qquad (8.11)$$

Then, from (8.10) and (8.11) we obtain

$$\tilde{f}_h'(x) - x\tilde{f}_h(x) = -\int_0^\infty \frac{d}{dt}\mathrm{E}\left(h\left(xe^{-t} + \sqrt{1 - e^{-2t}}X\right)\right)dt = h(x) - \mathrm{E}(h(X)).$$

Moreover, $\lim_{x \to \pm\infty} e^{-x^2/2}\tilde{f}_h(x) = 0$ because $\tilde{f}_h'$ has linear growth. By the uniqueness of the solution to Stein's equation, this implies that $\tilde{f}_h = f_h$, and the proof is complete. $\qquad\square$

The estimate (8.7) together with Propositions 8.1.6, 8.1.7, and 8.1.8 lead to the following Stein bounds for the total variation and the Kolmogorov and Wasserstein distances between an integrable random variable F and a random variable X with $N(0, 1)$ law:

$$d_{\mathrm{TV}}(F, X) \leq \sup_{f \in \mathscr{F}_{\mathrm{TV}}} |\mathrm{E}(f'(F) - Ff(F))|, \qquad (8.12)$$

$$d_{\mathrm{Kol}}(F, X) \leq \sup_{f \in \mathscr{F}_{\mathrm{Kol}}} |\mathrm{E}(f'(F) - Ff(F))|, \qquad (8.13)$$

$$d_{\mathrm{W}}(F, X) \leq \sup_{f \in \mathscr{F}_{\mathrm{W}}} |\mathrm{E}(f'(F) - Ff(F))|, \qquad (8.14)$$

where

$$\mathscr{F}_{\text{TV}} = \left\{ f \in C^1(\mathbb{R}) : \|f\|_\infty \leq \sqrt{\pi/2}, \|f'\|_\infty \leq 2 \right\},$$

$$\mathscr{F}_{\text{Kol}} = \left\{ f \in C^1(\mathbb{R}) : \|f\|_\infty \leq \sqrt{2\pi}/4, \|f'\|_\infty \leq 1 \right\},$$

$$\mathscr{F}_{\text{W}} = \left\{ f \in C^1(\mathbb{R}) : \|f'\|_\infty \leq \sqrt{2/\pi} \right\}.$$

Notice that for d_{TV} and d_{Kol} we can take the supremum over $f \in C^1$, because for any function h such that $\|h\|_\infty \leq 1$ we can find a sequence of continuous functions h_n, bounded by 1, such that h_n converges to h almost everywhere with respect to the measure $\ell + (\text{P} \circ F^{-1})$, where ℓ denotes the Lebesgue measure.

8.2 Stein Meets Malliavin

Suppose that again $B = (B_t)_{t \geq 0}$ is a Brownian motion defined on a probability space $(\Omega, \mathcal{F}, \text{P})$ such that $\mathcal{F}$ is generated by B. Combining Stein's method with Malliavin calculus leads to the following result, due to Nourdin and Peccati (2012).

Theorem 8.2.1 *Suppose that $F \in \mathbb{D}^{1,2}$ satisfies $F = \delta(u)$, where u belongs to $\text{Dom}\,\delta$. Let X be an $N(0, 1)$ random variable. Let $\mathscr{H}$ be a separating class of functions such that $\text{E}(|h(X)|) < \infty$ and $\text{E}(|h(F)|) < \infty$ for every $h \in \mathscr{H}$. Then,*

$$d_{\mathscr{H}}(F, X) \leq \sup_{h \in \mathscr{H}} \|f_h'\|_\infty \text{E}(|1 - \langle DF, u \rangle_H|).$$

Proof Using the duality relationship between D and δ (Proposition 3.2.3) and applying (5.8), we can write

$$|\text{E}(f_h'(F) - F f_h(F))| = |\text{E}(f_h'(F) - \langle D(f_h(F)), u \rangle_H)|$$
$$= |\text{E}(f_h'(F)(1 - \langle DF, u \rangle_H))|$$
$$\leq \sup_{h \in \mathscr{H}} \|f_h'\|_\infty \text{E}(|1 - \langle DF, u \rangle_H|),$$

which gives the estimate. $\qquad\qquad\square$

Now, applying the estimates (8.12)–(8.14), we obtain for $F = \delta(u) \in \mathbb{D}^{1,2}$ and X an $N(0, 1)$ random variable,

$$d_{\text{TV}}(F, X) \leq 2\text{E}(|1 - \langle DF, u \rangle_H|),$$
$$d_{\text{Kol}}(F, X) \leq \text{E}(|1 - \langle DF, u \rangle_H|),$$

and

$$d_W(F, X) \le \sqrt{\frac{2}{\pi}} E(|1 - \langle DF, u \rangle_H|).$$

Heuristically, this means that if $\langle DF, u \rangle_H$ is close to 1 in $L^1(\Omega)$ then the law of F is close to the $N(0, 1)$ law.

Possible choices of u for the representation $F = \delta(u)$ are given in Theorem 6.1.1 and Proposition 6.5.1.

Example 8.2.2 Suppose that $F = \int_0^T u_s dB_s$, where u is a progressively measurable process in $\mathbb{D}^{1,2}(H)$. Then

$$D_t F = u_t + \int_t^T D_t u_s dB_s$$

and

$$\langle u, DF \rangle_H = \|u\|_H^2 + \int_0^T \left(\int_t^T D_t u_s dB_s \right) u_t dt.$$

As a consequence,

$$d_{TV}(F, X) \le 2E(|1 - \|u\|_H^2|) + 2E\left(\left| \int_0^T \left(\int_t^T D_t u_s dB_s \right) u_t dt \right| \right)$$

$$\le 2E(|1 - \|u\|_H^2|) + 2\left(E\left(\int_0^T \left(\int_0^s u_t D_t u_s dt \right)^2 ds \right) \right)^{1/2}.$$

Therefore, a sequence $F_n = \int_0^T u_s^{(n)} dB_s$, where $u^{(n)} \in \mathbb{D}^{1,2}(H)$, converges in total variation to X, which has law $N(0, 1)$, if:

(i) $\|u^{(n)}\|_H^2 \to 1$ in $L^1(\Omega)$, and

(ii) $E \int_0^T \left(\int_0^s u_t^{(n)} D_t u_s^{(n)} dt \right)^2 ds \to 0$.

Now let $F \in \mathbb{D}^{1,2}$ be such that $E(F) = 0$. Let X be an $N(0, 1)$ random variable. Taking $u = -DL^{-1}F$, Theorem 8.2.1 gives

$$d_{\mathscr{H}}(F, X) \le \sup_{h \in \mathscr{H}} \|f_h'\|_\infty E(|1 - \langle DF, -DL^{-1}F \rangle_H|). \tag{8.15}$$

In some examples it is more convenient to compare the law of F with the law $N(0, \sigma^2)$, where $\sigma^2 > 0$. In that case, the previous results can be easily extended to the case where X has the $N(0, \sigma^2)$ law, and we obtain

$$d_{TV}(F, X) \le \frac{2}{\sigma^2} E(|\sigma^2 - \langle DF, u \rangle_H|), \qquad (8.16)$$

$$d_{Kol}(F, X) \le \frac{1}{\sigma^2} E(|\sigma^2 - \langle DF, u \rangle_H|),$$

$$d_W(F, X) \le \frac{1}{\sigma^2} \sqrt{\frac{2}{\pi}} E(|\sigma^2 - \langle DF, u \rangle_H|).$$

8.3 Normal Approximation on a Fixed Wiener Chaos

In this section we will focus on the normal approximation when the random variable F belongs to a fixed Wiener chaos of order $q \ge 2$. To simplify the presentation we will consider only the total variation distance. First, we nprove the following result.

Proposition 8.3.1 *Suppose that $F \in \mathcal{H}_q$ for some $q \ge 2$, and $E(F^2) = \sigma^2$. Let X be an $N(0, \sigma^2)$ random variable. Then*

$$d_{TV}(F, X) \le \frac{2}{q\sigma^2} \sqrt{\mathrm{Var}(\|DF\|_H^2)}.$$

Proof We have $L^{-1}F = -(1/q)F$. Taking into account that $E(\|DF\|_H^2) = q\sigma^2$, we obtain

$$E(|\sigma^2 - \langle DF, -DL^{-1}F \rangle_H|) = E\left(\left|\sigma^2 - \frac{1}{q}\|DF\|_H^2\right|\right) \le \frac{1}{q} \sqrt{\mathrm{Var}(\|DF\|_H^2)}.$$

Then the desired estimate follows from (8.16). □

The next proposition shows that the variance of $\|DF\|_H^2$ is equivalent to $E(F^4) - 3\sigma^4$.

Proposition 8.3.2 *Suppose that $F = I_q(f) \in \mathcal{H}_q$, $q \ge 2$, and $E(F^2) = \sigma^2$. Then*

$$\mathrm{Var}(\|DF\|_H^2) \le \frac{(q-1)q}{3}(E(F^4) - 3\sigma^4) \le (q-1)\mathrm{Var}(\|DF\|_H^2).$$

Proof The proof will be done in several steps. First we will derive two formulas for the variance of $\|DF\|_H^2$ and for $E(F^4) - 3\sigma^4$.

Step 1 We claim that

$$\mathrm{Var}(\|DF\|_H^2) = \sum_{r=1}^{q-1} r^2 (r!)^2 \binom{q}{r}^4 (2q - 2r)! \|f \tilde{\otimes}_r f\|_{H^{\otimes(2q-2r)}}^2. \qquad (8.17)$$

In fact, applying Proposition 4.2.1, we have $D_t F = q I_{q-1}(f(\cdot, t))$. Then, using the product formula for multiple stochastic integrals (see (4.2)), we obtain

$$
\begin{aligned}
\|DF\|_H^2 &= q^2 \int_0^\infty I_{q-1}(f(\cdot, t))^2 \, dt \\
&= q^2 \sum_{r=0}^{q-1} r! \binom{q-1}{r}^2 I_{2q-2r-2}(f \tilde{\otimes}_{r+1} f) \\
&= q^2 \sum_{r=1}^{q} (r-1)! \binom{q-1}{r-1}^2 I_{2q-2r}(f \tilde{\otimes}_r f) \\
&= qq! \|f\|_{H^{\otimes q}}^2 + q^2 \sum_{r=1}^{q-1} (r-1)! \binom{q-1}{r-1}^2 I_{2q-2r}(f \tilde{\otimes}_r f). \quad (8.18)
\end{aligned}
$$

Taking into account that $E(\|DF\|_H^2) = qq! \|f\|_{H^{\otimes q}}^2$, and using the isometry property of multiple integrals (4.1), we show formula (8.17).

Step 2   We claim that

$$
E(F^4) - 3\sigma^4 = \frac{3}{q} \sum_{r=1}^{q-1} r(r!)^2 \binom{q}{r}^4 (2q - 2r)! \|f \tilde{\otimes}_r f\|_{H^{\otimes(2q-2r)}}^2. \quad (8.19)
$$

Indeed, using that $-L^{-1}F = (1/q)F$ and $L = -\delta D$ we first write

$$
\begin{aligned}
E(F^4) = E(F \times F^3) &= E((-\delta D L^{-1} F)F^3) = E(\langle -DL^{-1}F, D(F^3) \rangle_H) \\
&= \frac{1}{q} E(\langle DF, D(F^3) \rangle_H) = \frac{3}{q} E(F^2 \|DF\|_H^2). \quad (8.20)
\end{aligned}
$$

By the product formula for multiple integrals (4.2),

$$
F^2 = I_q(f)^2 = q! \|f\|_{H^{\otimes q}}^2 + \sum_{r=1}^{q} r! \binom{q}{r}^2 I_{2q-2r}(f \tilde{\otimes}_r f). \quad (8.21)
$$

Using the isometry property for multiple integrals (4.1), and expressions (8.18) and (8.21), we can compute $E(F^2 \|DF\|_H^2)$; substituting its value into (8.20), we get formula (8.19).

Step 3   Comparing formulas (8.17) and (8.19) yields the desired estimates.   □

In summary, we have proved the following result.

Proposition 8.3.3 *Let $q \geq 2$ and $F = I_q(f)$. Set $E(F^2) = \sigma^2 > 0$. Then, if X is a random variable with law $N(0, \sigma^2)$,*

$$d_{\mathrm{TV}}(F, X) \leq \frac{2}{\sigma^2 q} \sqrt{\mathrm{Var}(\|DF\|_H^2)} \leq \frac{2}{\sigma^2} \sqrt{\frac{q-1}{3q}} (E(F^4) - 3\sigma^4).$$

These results can be applied to derive the so-called *fourth-moment the-orem*, proved by Nualart and Peccati (2005) (see also Nualart and Ortiz-Latorre, 2007), which represents a drastic simplification of the method of moments.

Theorem 8.3.4 *Fix $q \geq 2$. Consider a sequence of multiple stochastic integrals of order q, $F_n = I_q(f_n) \in \mathcal{H}_q$, $n \geq 1$, such that*

$$\lim_{n \to \infty} E(F_n^2) = \sigma^2.$$

The following conditions are equivalent:

(i) $F_n \xrightarrow{\mathcal{L}} N(0, \sigma^2)$ *as $n \to \infty$;*
(ii) $E(F_n^4) \to 3\sigma^4$ *as $n \to \infty$;*
(iii) $\|DF_n\|_H^2 \to q\sigma^2$ *in $L^2(\Omega)$ as $n \to \infty$;*
(iv) *for all $1 \leq r \leq q - 1$, $f_n \otimes_r f_n \to 0$ as $n \to \infty$.*

Proof First notice that (i) implies (ii) because, for any $p > 2$, the hy-percontractivity property of the Ornstein–Uhlenbeck semigroup (see (5.3)) implies that

$$\sup_n \|F_n\|_p \leq (p-1)^{q/2} \sup_n \|F_n\|_2 < \infty.$$

The equivalence of (ii) and (iii) follows from Proposition 8.3.2, and these two conditions imply (i), with convergence in total variation. Moreover, (iv) implies (ii) and (iii), in view of formulas (8.17) and (8.19), because

$$\|f_n \tilde{\otimes}_r f_n\|_{H^{\otimes(2q-2r)}} \leq \|f_n \otimes_r f_n\|_{H^{\otimes(2q-2r)}}.$$

Finally, it remains to show that (ii) implies (iv). From (8.21), using the isometry property of multiple stochastic integrals, we obtain

$$\begin{aligned}
E(F_n^4) &= \sum_{r=0}^{q} (r!)^2 \binom{q}{r}^2 (2q - 2r)! \|f_n \tilde{\otimes}_r f_n\|_{H^{\otimes(2q-2r)}}^2 \\
&= (q!)^2 \|f_n\|_{H^{\otimes q}}^4 + (2q)! \|f_n \tilde{\otimes} f_n\|_{H^{\otimes 2q}}^2 \\
&\quad + \sum_{r=1}^{q-1} (r!)^2 \binom{q}{r}^2 (2q - 2r)! \|f_n \tilde{\otimes}_r f_n\|_{H^{\otimes(2q-2r)}}^2.
\end{aligned}$$

Then one can show that $(2q)!\|f_n \tilde{\otimes} f_n\|^2_{H^{\otimes 2q}}$ equals $2(q!)^2\|f_n\|^4_{H^{\otimes q}}$ plus a linear combination of the terms $\|f_n \otimes_r f_n\|^2_{H^{\otimes(2q-2r)}}$, where $1 \le r \le q-1$, with strictly positive coefficients. Therefore, $E(F_n^4)$ can be expressed as $3(q!)^2\|f_n\|^4_{H^{\otimes q}} = 3\sigma^4$ plus a linear combination with strictly positive coefficients of contractions $\|f_n \otimes_r f_n\|^2_{H^{\otimes(2q-2r)}}$ and symmetric contractions $\|f_n \tilde{\otimes}_r f_n\|^2_{H^{\otimes(2q-2r)}}$, where $1 \le r \le q-1$. By (ii), all these contractions must converge to zero. This completes the proof. $\qquad\square$

The following multidimensional extension of the fourth-moment theorem was proved in Peccati and Tudor (2005). Notice that the convergence of the marginal distributions to normal random variables implies automatically the joint convergence to a Gaussian random vector with independent components.

Theorem 8.3.5 *Let $d \ge 2$ and $1 \le q_1 < \cdots < q_d$. Consider a sequence of random vectors whose components are multiple stochastic integrals of orders $q_1, \ldots, q_d$; that is,*

$$F_n = (F_n^1, \ldots, F_n^d) = (I_{q_1}(f_n^1), \ldots, I_{q_d}(f_n^d)), \quad n \ge 1,$$

where $f_n^i \in L_s^2(\mathbb{R}^{q_i})$. Suppose that, for any $1 \le i \le d$,

$$\lim_{n \to \infty} E((F_n^i)^2) = \sigma_i^2. \tag{8.22}$$

Then, the following two conditions are equivalent:

(i) *$F_n \xrightarrow{\mathcal{L}} N_d(0, \Sigma)$, where Σ is a $d \times d$ diagonal matrix such that $\Sigma_{ii} = \sigma_i^2$ for each $i = 1, \ldots, d$;*

(ii) *for every $i = 1, \ldots, d$, $F_n^i \xrightarrow{\mathcal{L}} N(0, \sigma_i^2)$.*

Proof It suffices to show that (ii) implies (i). The sequence of random vectors $(F_n)_{n \ge 1}$ is tight by condition (8.22). Then it suffices to show that the limit in distribution of any converging subsequence is $N_d(0, \Sigma)$. We can then assume that the sequence $(F_n)_{n \ge 1}$ converges in law to some random vector F_∞, and it only remains to show that the law of F_∞ is $N_d(0, \Sigma)$. For every n and $t \in \mathbb{R}^d$, define $\varphi_n(t) = E(e^{i\langle t, F_n \rangle})$. Let $\varphi(t)$ be the characteristic function of the random vector F_∞. We know that, for all $j = 1, \ldots, d$,

$$\frac{\partial \varphi_n}{\partial t_j}(t) = iE\left(F_n^j e^{i\langle t, F_n \rangle}\right) \xrightarrow[n \to \infty]{} iE\left(F_\infty^j e^{i\langle t, F_\infty \rangle}\right) = \frac{\partial \varphi}{\partial t_j}(t). \tag{8.23}$$

Moreover, using the definition of the operator L, Proposition 5.2.1, and the duality relation between D and δ we have

$$\mathrm{E}\left(F_n^j e^{i\langle t, F_n\rangle}\right) = -\frac{1}{q_j}\mathrm{E}\left(LF_n^j e^{i\langle t, F_n\rangle}\right) = -\frac{1}{q_j}\mathrm{E}\left(-\delta D(F_n^j)e^{i\langle t, F_n\rangle}\right)$$

$$= \frac{1}{q_j}\mathrm{E}\left(\langle DF_n^j, D(e^{i\langle t, F_n\rangle})\rangle_H\right) = \frac{i}{q_j}\sum_{h=1}^{d} t_h \mathrm{E}\left(e^{i\langle t, F_n\rangle}\gamma_{F_n}^{jh}\right),$$

where γ_{F_n} is the Malliavin matrix of the random vector F_n, introduced in (7.5). Therefore,

$$\frac{\partial \varphi_n}{\partial t_j}(t) = -\frac{1}{q_j}\sum_{h=1}^{d} t_h \mathrm{E}\left(e^{i\langle t, F_n\rangle}\gamma_{F_n}^{jh}\right). \tag{8.24}$$

We claim that the following convergences hold in $L^2(\Omega)$ for all $1 \le j, h \le d$, as n tends to infinity:

$$\gamma_{F_n}^{jh} \to 0, \quad j \ne h, \tag{8.25}$$

and

$$\gamma_{F_n}^{jj} \to q_j\sigma_j^2. \tag{8.26}$$

In fact, (8.26) follows from condition (iii) in Theorem 8.3.4 and the fact that $\gamma_{F_n}^{jj} = \|DF_n^j\|_H^2$. Furthermore, using the product formula for multiple stochastic integrals and assuming that $j < h$ yields

$$\mathrm{E}\left(\langle DF_n^j, DF_n^h\rangle_H^2\right) = q_j^2 q_h^2 \mathrm{E}\left(\left(\int_0^\infty I_{q_j-1}(f_n^j(\cdot, t))I_{q_h-1}(f_n^h(\cdot, t))dt\right)^2\right)$$

$$= q_j^2 q_h^2 \mathrm{E}\left(\left(\sum_{r=0}^{q_j-1}\binom{q_j-1}{r}\binom{q_h-1}{r}r! I_{q_j+q_h-2-2r}(f_n^j \otimes_{r+1} f_n^h)\right)^2\right)$$

$$= q_j^2 q_h^2 \sum_{r=0}^{q_j-1}\binom{q_j-1}{r}^2\binom{q_h-1}{r}^2(r!)^2(q_j+q_h-2-2r)!\|f_n^j\tilde\otimes_{r+1}f_n^h\|_{H^{\otimes(q_j+q_h-2-2r)}}^2$$

$$\le q_j^2 q_h^2 \sum_{r=1}^{q_j}\binom{q_j-1}{r-1}^2\binom{q_h-1}{r-1}^2((r-1)!)^2(q_j+q_h-2r)!\|f_n^j\otimes_r f_n^h\|_{H^{\otimes(q_j+q_h-2r)}}^2.$$

Then, to show (8.25) it suffices to check that $\|f_n^j \otimes_r f_n^h\|_{H^{\otimes(q_j+q_h-2r)}}^2$ converges to zero for all $1 \le r \le q_j$. This follows from

$$\|f_n^j \otimes_r f_n^h\|_{H^{\otimes(q_j+q_h-2r)}}^2 = \langle f_n^j \otimes_{q_j-r} f_n^j, f_n^h \otimes_{q_h-r} f_n^h\rangle_{H^{\otimes 2r}}$$

$$\le \|f_n^j \otimes_{q_j-r} f_n^j\|_{H^{\otimes 2r}}\|f_n^h \otimes_{q_h-r} f_n^h\|_{H^{\otimes 2r}}$$

and the fact that $\|f_n^h \otimes_{q_h-r} f_n^h\|_{H^{\otimes 2r}}$ converges to zero by condition (iv) in Theorem 8.3.4, because $1 \le q_h - r \le q_h - 1$. Finally, (8.25) and (8.26) allow us to take the limit as $n \to \infty$ on the right-hand side of (8.24) and, in view of (8.23), we obtain

$$\frac{\partial \varphi}{\partial t_j}(t) = -t_j \sigma_j^2 \varphi(t),$$

for $j = 1, \ldots, d$. As a consequence, φ is the characteristic function of the law $N_d(0, \Sigma)$. $\qquad \square$

8.4 Chaotic Central Limit Theorem

When the random variables are not in a fixed chaos, we can establish the following result proved in Hu and Nualart (2005).

Theorem 8.4.1 *Consider a sequence of centered square integrable random variables* $(F_n)_{n \geq 1}$ *with Wiener chaos expansions* $F_n = \sum_{q=1}^{\infty} I_q(f_{q,n})$. *Suppose that:*

(i) *for all* $q \geq 1$, $\lim_{n \to \infty} q! \|f_{q,n}\|_{H^{\otimes q}}^2 = \sigma_q^2$;
(ii) *for all* $q \geq 2$ *and* $1 \leq r \leq q - 1$, $f_{q,n} \otimes_r f_{q,n} \to 0$ *as* $n \to \infty$;
(iii) $\lim_{N \to \infty} \limsup_{n \to \infty} \sum_{q=N+1}^{\infty} q! \|f_{q,n}\|_{H^{\otimes q}}^2 = 0$.

Then, as n tends to infinity,

$$F_n \xrightarrow{\mathcal{L}} N(0, \sigma^2),$$

where $\sigma^2 = \sum_{q=1}^{\infty} \sigma_q^2$.

Proof Let ξ_q, $q \geq 1$, be independent centered Gaussian random variables with variances σ_q^2. For every $N \geq 1$, set $F_n^N = \sum_{q=1}^{N} I_q(f_{q,n})$ and $\xi^N = \sum_{q=1}^{N} \xi_q$. By Theorems 8.3.4 and 8.3.5, for each fixed N we have that F_n^N converges in law to ξ^N as n tends to infinity. Define also $\xi = \sum_{q=1}^{\infty} \xi_q$. Let f be a $C_b^1(\mathbb{R})$ function such that $\|f\|_\infty$ and $\|f'\|_\infty$ are bounded by one. Then

$$|E(f(F_n)) - E(f(\xi))| \leq |E(f(F_n)) - E(f(F_n^N))|$$
$$+ |E(f(F_n^N)) - E(f(\xi^N))| + |E(f(\xi^N)) - E(f(\xi))|$$
$$\leq \left(\sum_{q=N+1}^{\infty} q! \|f_{q,n}\|_{H^{\otimes q}}^2 \right)^{1/2} + |E(f(F_n^N)) - E(f(\xi^N))|$$
$$+ |E(f(\xi^N)) - E(f(\xi))|.$$

Taking first the limit as n tends to infinity, and then the limit as N tends to infinity, and applying conditions (i) and (iii), we finish the proof. $\qquad \square$

Taking into account Theorem 8.3.5 and assuming condition (i), condition (ii) is equivalent to

(iia) $\lim_{n\to\infty} E(I_q(f_{q,n})^4) = 3\sigma_q^4$, $q \geq 2$,

  or

(iib) $\lim_{n\to\infty} \|D(I_q(f_{q,n}))\|_H^2 \to q\sigma_q^2$ in $L^2(\Omega)$, $q \geq 2$.

Theorem 8.4.1 implies the convergence in law of the whole sequence $(I_q(f_{q,n}))_{q\geq 1}$ to an infinite-dimensional Gaussian vector with independent components, and it can be regarded as a chaotic central limit theorem.

As an application we are going to present a simple proof of the classical Breuer–Major theorem (see Breuer and Major, 1983). A function $f \in L^2(\mathbb{R}, \gamma)$, where $\gamma = N(0, 1)$, has *Hermite rank* $d \geq 1$ if f has a series expansion in $L^2(\mathbb{R}, \gamma)$ of the form

$$f(x) = \sum_{q=d}^{\infty} a_q h_q(x),$$

where h_q is the qth Hermite polynomial, introduced in (4.3), and $a_d \neq 0$. For instance, $f(x) = |x|^p - \int_{\mathbb{R}} |x|^p d\gamma(x)$ has Hermite rank 2 if $p \geq 1$.

Let $Y = (Y_k)_{k\in\mathbb{Z}}$ be a centered Gaussian stationary sequence with unit variance. Set $\rho(v) = E(Y_0 Y_v)$ for $v \in \mathbb{Z}$.

Theorem 8.4.2 *Let $f \in L^2(\mathbb{R}, \gamma)$ with Hermite rank d and assume that*

$$\sum_{v\in\mathbb{Z}} |\rho(v)|^d < \infty.$$

Then

$$V_n := \frac{1}{\sqrt{n}} \sum_{k=1}^{n} f(Y_k) \xrightarrow{\mathcal{L}} N(0, \sigma^2)$$

as $n \to \infty$, where $\sigma^2 = \sum_{q=d}^{\infty} q! a_q^2 \sum_{v\in\mathbb{Z}} \rho(v)^q$.

Proof There exists a sequence $(e_k)_{k\geq 1}$ in $H = L^2([0, \infty))$ such that

$$\langle e_k, e_j \rangle_H = \rho(k - j).$$

The sequence $(B(e_k))_{k\geq 1}$ has the same law as $(Y_k)_{k\geq 1}$, and we can replace V_n by

$$G_n = \frac{1}{\sqrt{n}} \sum_{k=1}^{n} \sum_{q=d}^{\infty} a_q h_q(B(e_k)) = \sum_{q=d}^{\infty} I_q(f_{q,n}),$$

where

$$f_{q,n} = \frac{a_q}{\sqrt{n}} \sum_{k=1}^{n} e_k^{\otimes q}.$$

Then it suffices to show that the kernels $f_{q,n}$ satisfy conditions (i), (ii), and (iii) of Theorem 8.4.1 with $\sigma_q^2 = q! a_q^2 \sum_{v \in \mathbb{Z}} \rho(v)^q$.

Step 1 First we show condition (i). We can write

$$q! \|f_{n,q}\|_{H^{\otimes q}}^2 = \frac{q! a_q^2}{n} \sum_{i,j=1}^{n} \rho(i-j)^q = q! a_q^2 \sum_{v \in \mathbb{Z}} \rho(v)^q \left(1 - \frac{|v|}{n}\right) \mathbf{1}_{\{|v| < n\}} \qquad (8.27)$$

and, by the dominated convergence theorem,

$$E(F_n^2) = q! \|f_{n,q}\|_{H^{\otimes q}}^2 \to q! a_q^2 \sum_{v \in \mathbb{Z}} \rho(v)^q = \sigma_q^2$$

as n tends to infinity, which implies condition (i).

Step 2 We need to show that, for $r = 1, \dots, q-1$ and $q \geq 2$,

$$f_{q,n} \otimes_r f_{q,n} = \frac{a_q^2}{n} \sum_{k,j=1}^{n} \rho(k-j)^r e_k^{\otimes(q-r)} \otimes e_j^{\otimes(q-r)} \to 0.$$

We have

$$\|f_{q,n} \otimes_r f_{q,n}\|_{H^{\otimes(2q-2r)}}^2 = \frac{a_q^4}{n^2} \sum_{i,j,k,\ell=1}^{n} \rho(k-j)^r \rho(i-\ell)^r \rho(k-i)^{q-r} \rho(j-\ell)^{q-r}.$$

Using $|\rho(k-j)^r \rho(k-i)^{q-r}| \leq |\rho(k-j)|^q + |\rho(k-i)|^q$, we obtain

$$\|f_{q,n} \otimes_r f_{q,n}\|_{H^{\otimes(2q-2r)}}^2 \leq 2 a_q^4 \sum_{k \in \mathbb{Z}} |\rho(k)|^q \left(n^{-1+r/q} \sum_{|i| \leq n} |\rho(i)|^r\right)$$
$$\times \left(n^{-1+(q-r)/q} \sum_{|j| \leq n} |\rho(j)|^{q-r}\right).$$

Then it remains to show that, for $r = 1, \dots, q-1$,

$$n^{-1+r/q} \sum_{|i| \leq n} |\rho(i)|^r \to 0.$$

This follows from Hölder's inequality. Indeed, for a fixed $\delta \in (0, 1)$, we have the estimates

$$n^{-1+r/q} \sum_{|i| \leq [n\delta]} |\rho(i)|^r \leq n^{-1+r/q} (2[n\delta] + 1)^{1-r/q} \left(\sum_{i \in \mathbb{Z}} |\rho(i)|^q\right)^{r/q} \leq c \delta^{1-r/q}$$

and

$$n^{-1+r/q} \sum_{[n\delta]<|i|\leq n} |\rho(i)|^r \leq \left(\sum_{[n\delta]<|i|\leq n} |\rho(i)|^q \right)^{r/q}.$$

The first term converges to zero as δ tends to zero and the second converges to zero for fixed δ as $n \to \infty$.

Step 3   Finally, condition (iii) is an immediate consequence of (8.27). This concludes the proof. $\square$

8.5 Applications to Fractional Brownian Motion

In this section we apply the theorems established in the previous sections to derive several results on the asymptotic behavior of functionals of fractional Brownian motion. Fractional Brownian motion $B^H = (B_t^H)_{t\in\mathbb{R}}$ is a mean-zero Gaussian process with covariance given by (see (3.7) for the case $t \geq 0$):

$$E(B_s^H B_t^H) = \tfrac{1}{2}\left(|s|^{2H} + |t|^{2H} - |t-s|^{2H}\right), \quad s, t \in \mathbb{R},$$

where $H \in (0,1)$ is the Hurst parameter.

Fractional Brownian motion B^H has stationary increments. More precisely, the sequence $Y_n^H = B_{n+1}^H - B_n^H$, $n \in \mathbb{Z}$ (called fractional noise), is Gaussian, stationary, with mean zero, unit variance, and covariance given by

$$\rho_H(n) = E(Y_0^H Y_n^H) = \tfrac{1}{2}\left(|n+1|^{2H} + |n-1|^{2H} - 2|n|^{2H}\right). \tag{8.28}$$

In the next proposition we show that fractional Brownian motion has finite $1/H$-variation. For $q > 0$ we put $c_q = E(|X|^q)$, X being an $N(0,1)$ random variable.

Proposition 8.5.1   *Let B^H be a fractional Brownian motion with Hurst parameter H. Fix $T > 0$ and set $t_i = iT/n$ for $1 \leq i \leq n$. Define $\Delta B_{t_i}^H = B_{t_i}^H - B_{t_{i-1}}^H$. Then, as $n \to \infty$,*

$$\sum_{i=1}^n |\Delta B_{t_i}^H|^{1/H} \xrightarrow{L^2(\Omega),a.s.} c_{1/H} T.$$

Proof   By the self-similarity of fractional Brownian motion, the random variable $\sum_{i=1}^n |\Delta B_{t_i}^H|^{1/H}$ has the same law as

$$\frac{T}{n} \sum_{i=1}^n |B_i^H - B_{i-1}^H|^{1/H}.$$

The sequence $(Y_i^H = B_i^H - B_{i-1}^H)_{i \geq 1}$ is stationary and ergodic. Therefore, the ergodic theorem implies the desired convergence. □

The covariance (8.28) satisfies $\rho_H(|n|) \sim H(2H-1)|n|^{2H-2}$ as $|n| \to \infty$. This implies that, for any integer $q \geq 2$ such that $H < 1 - 1/(2q)$, we have

$$\sum_{v \in \mathbb{Z}} |\rho_H(v)|^q < \infty.$$

As a consequence, Theorem 8.4.2 implies the following asymptotic result for the Hermite variations:

$$\frac{1}{\sqrt{n}} \sum_{k=1}^{n} h_q(B_k^H - B_{k-1}^H) \overset{\mathcal{L}}{\longrightarrow} N(0, \sigma_{H,q}^2),$$

for each $q \geq 2$, where $\sigma_{H,q}^2 = q! \sum_{v \in \mathbb{Z}} \rho_H(v)^q$.

More generally, the Breuer–Major theorem (Theorem 8.4.2) leads to the following convergence, taking into account that $|x|^q - c_q$ has Hermite rank 1 when $q > 0$ is not an even integer.

Theorem 8.5.2 *Suppose that* $H < \frac{1}{2}$ *and* $q \geq 1$ *is not an even integer. As* $n \to \infty$, *we have*

$$\frac{1}{\sqrt{n}} \sum_{k=1}^{n} \left(n^{qH} |\Delta_{k,n} B^H|^q - c_q \right) \overset{\mathcal{L}}{\longrightarrow} N(0, \tilde{\sigma}_{H,q}^2),$$

where $\Delta_{k,n} B^H = B_{k/n}^H - B_{(k-1)/n}^H$ *and* $\tilde{\sigma}_{H,q}^2 = \sum_{q=1}^{\infty} q! a_q^2 \sum_{v \in \mathbb{Z}} \rho_H(v)^q$.

In the case of a quadratic variation, Proposition 8.3.1 allows us to deduce the rate of convergence in the total variation distance. Define, for $n \geq 1$,

$$S_n = \sum_{k=1}^{n} (\Delta_{k,n} B^H)^2.$$

Then, from Proposition 8.5.1,

$$n^{2H-1} S_n \overset{L^2(\Omega), \text{a.s.}}{\longrightarrow} 1$$

as n tends to infinity. To study the asymptotic normality associated with this almost sure convergence, consider

$$F_n = \frac{1}{\sigma_n} \sum_{k=1}^{n} \left(n^{2H} (\Delta_{k,n} B^H)^2 - 1 \right) \overset{d}{=} \frac{1}{\sigma_n} \sum_{k=1}^{n} \left((B_k^H - B_{k-1}^H)^2 - 1 \right),$$

where σ_n is such that $E(F_n^2) = 1$. With this notation, we have the following result.

Theorem 8.5.3 *Assume* $H < \frac{3}{4}$ *and let* X *be an* $N(0, 1)$ *random variable. Then,* $\lim_{n\to\infty} \sigma_n^2/n = 2\sum_{r\in\mathbb{Z}} \rho_H^2(r)$ *and*

$$d_{TV}(F_n, X) \leq c_H \times \begin{cases} n^{-1/2} & \text{if } H \in (0, \frac{5}{8}), \\ n^{-1/2}(\log n)^{3/2} & \text{if } H = \frac{5}{8}, \\ n^{4H-3} & \text{if } H \in (\frac{5}{8}, \frac{3}{4}). \end{cases}$$

Proof There exists a sequence $(e_k)_{k\geq 1}$ in $L^2([0, \infty))$ such that

$$\langle e_k, e_j \rangle_H = \rho_H(k - j).$$

The sequence $(B(e_k))_{k\geq 1}$ has the same law as $(B_k^H - B_{k-1}^H)_{k\geq 1}$, and we may replace F_n by

$$G_n = \frac{1}{\sigma_n} \sum_{k=1}^{n} (B(e_k)^2 - 1) = I_2(f_n),$$

where

$$f_n = \frac{1}{\sigma_n} \sum_{k=1}^{n} e_k \otimes e_k.$$

By the isometry property of multiple integrals,

$$1 = E(G_n^2) = 2\|f_n\|_{L^2([0,\infty)^2)}^2 = \frac{2}{\sigma_n^2} \sum_{k,j=1}^{n} \rho_H^2(k - j) = \frac{2n}{\sigma_n^2} \sum_{|r|<n} \left(1 - \frac{|r|}{n}\right) \rho_H^2(r).$$

Since $\sum_{r\in\mathbb{Z}} \rho_H^2(r) < \infty$, because $H < \frac{3}{4}$ we can deduce that $\lim_{n\to\infty} \sigma_n^2/n = 2\sum_{r\in\mathbb{Z}} \rho_H^2(r)$.

We can write $D_r(I_2(f_n)) = 2I_1(f_n(\cdot, r))$ and

$$\|D(I_2(f_n))\|_H^2 = 4\left(I_2(f_n \otimes_1 f_n) + \|f_n\|_H^2\right) = 4I_2(f_n \otimes_1 f_n) + 2.$$

Therefore

$$\begin{aligned} \text{Var}\left(\|D(I_2(f_n))\|_H^2\right) &= 16E\left((I_2(f_n \otimes_1 f_n))^2\right) \\ &= 32\|f_n \otimes_1 f_n\|_{L^2([0,\infty)^2)}^2 \\ &= \frac{32}{\sigma_n^4} \sum_{k,j,i,\ell=1}^{n} \rho_H(k - j)\rho_H(i - \ell)\rho_H(k - i)\rho_H(j - \ell) \\ &\leq \frac{32}{\sigma_n^4} \sum_{i,\ell=1}^{n} (\rho_n * \rho_n)(i - \ell)^2 \\ &\leq \frac{32n}{\sigma_n^4} \sum_{k\in\mathbb{Z}} (\rho_n * \rho_n)(k)^2 = \frac{32n}{\sigma_n^4}\|\rho_n * \rho_n\|_{\ell^2(\mathbb{Z})}^2, \end{aligned}$$

where $\rho_n(k) = |\rho_H(k)|\mathbf{1}_{\{|k|\leq n-1\}}$. Applying Young's inequality yields

$$\|\rho_n * \rho_n\|^2_{\ell^2(\mathbb{Z})} \leq \|\rho_n\|^4_{\ell^{4/3}(\mathbb{Z})},$$

so that

$$\mathrm{Var}\left(\|D(I_2(f_n))\|^2_H\right) \leq \frac{32n}{\sigma_n^4}\left(\sum_{|k|<n}|\rho_H(k)|^{4/3}\right)^3.$$

Thus,

$$d_{\mathrm{TV}}(F_n, X) \leq \frac{4\sqrt{2n}}{\sigma_n^2}\left(\sum_{|k|<n}|\rho_H(k)|^{4/3}\right)^{3/2},$$

and the result follows from $\rho_H(k) \sim H(2H-1)|k|^{2H-2}$ as $|k| \to \infty$. $\qquad\square$

As a consequence, we have established the following convergence in law:

$$\sqrt{n}(n^{2H-1}S_n - 1) \xrightarrow{\mathcal{L}} N\left(0, 2\sum_{r\in\mathbb{Z}}\rho^2(r)\right).$$

These results on the quadratic variation can be applied to the estimation of the Hurst parameter. Consider the estimator of H from the observations $B^H_{k/n}$, $1 \leq k \leq n$, given by

$$\hat{H}_n = \frac{1}{2} - \frac{\log S_n}{2\log n}.$$

Then, $\hat{H}_n$ is strongly consistent; that is, it satisfies $\hat{H}_n \xrightarrow{\text{a.s.}} H$ and is asymptotically normal. In other words,

$$\sqrt{n}\log n(\hat{H}_n - H) \xrightarrow{\mathcal{L}} N\left(0, \frac{1}{2}\sum_{r\in\mathbb{Z}}\rho^2_H(r)\right).$$

In Nourdin and Peccati (2015) the following optimal version of the fourth-moment theorem was derived.

Theorem 8.5.4 *Consider a sequence of random variables $(F_n)_{n\geq 1}$ in the chaos $\mathcal{H}_q$, where $q \geq 2$. Suppose that $E(F_n^2) = 1$. Let X be an $N(0,1)$ random variable. Then there exist positive constants c and C such that*

$$cM(F_n) \leq d_{\mathrm{TV}}(F_n, X) \leq CM(F_n),$$

where $M(F_n) = \max(|E(F_n^3)|, E(F_n^4) - 3)$.

As a an application of Theorem 8.5.4, it was shown in Biermé *et al.* (2012) that the sequence $F_n = \sigma_n^{-1} \sum_{k=1}^{n} \left((B_k^H - B_{k-1}^H)^2 - 1 \right)$ satisfies

$$d_{TV}(F_n, X) \leq c_H \times \begin{cases} n^{-1/2} & \text{if } H \in (0, \frac{2}{3}), \\ n^{-1/2}(\log n)^2 & \text{if } H = \frac{2}{3}, \\ n^{6H-9/2} & \text{if } H \in (\frac{2}{3}, \frac{3}{4}). \end{cases}$$

8.6 Convergence of Densities

The total variation distance between the laws of two absolutely continuous random variables F and G is equivalent to the $L^1(\mathbb{R})$ distance of the densities, and we have the identity

$$d_{TV}(F, G) = \int_{\mathbb{R}} |p_F(x) - p_G(x)| dx,$$

where p_F and p_G are the densities of the random variables F and G, respectively. Therefore, Proposition 8.3.3 implies the convergence in $L^1(\mathbb{R})$ of the density of a random variable in a fixed chaos to the standard Gaussian density. Under nondegeneracy conditions on the norm of the Malliavin derivative (similar to the conditions imposed to ensure the existence and regularity of densities), one can estimate the uniform distance between the densities. This is the content of the next result, proved by Hu *et al.* (2014b).

Theorem 8.6.1 *Let $F \in \mathcal{H}_q$, $q \geq 2$, such that $E(F^2) = 1$ and $E(\|DF\|_H^{-6}) \leq M$. Then*

$$\sup_{x \in \mathbb{R}} |p_F(x) - \phi(x)| \leq C_{M,q} \sqrt{E(F^4) - 3},$$

where ϕ is the density of the law $N(0, 1)$.

Using the notion of the *Fisher information*, Nourdin and Nualart (2016) provided a proof of this theorem under the weaker assumption $E(\|DF\|_H^{-4-\epsilon}) \leq M$ for some $\epsilon > 0$.

Proof of Theorem 8.6.1 Using the density formula (7.1), property (3.5), and $\delta DF = qF$, we can write

$$p_F(x) = E\left(1_{\{F>x\}} \delta\left(\frac{DF}{\|DF\|_H^2}\right)\right)$$

$$= E\left(1_{\{F>x\}} \frac{qF}{\|DF\|_H^2}\right) - E\left(1_{\{F>x\}} \langle DF, D(\|DF\|_H^{-2})\rangle_H\right)$$

$$= E(1_{\{F>x\}}F) + E\left(q\|DF\|_H^{-2} - 1\right) - E\left(1_{\{F>x\}} \langle DF, D(\|DF\|_H^{-2})\rangle_H\right).$$

The quantities $E(|q\|DF\|_H^{-2} - 1|)$ and $E(|\langle DF, D(\|DF\|_H^{-2})\rangle_H|)$ can be estimated by a constant times $\sqrt{E(F^4) - 3}$. Taking into account that

$$\phi(x) = E(\mathbf{1}_{\{X>x\}}X),$$

where X has the $N(0, 1)$ distribution, it suffices to estimate the difference

$$E(\mathbf{1}_{\{F>x\}}F) - E(\mathbf{1}_{\{X>x\}}X),$$

which can be done by Stein's method and the Malliavin calculus. We omit the details of the proof of these two estimations. □

Example 8.6.2 Let $q = 2$ and $F = \sum_{i=1}^{\infty} \lambda_i(B(e_i)^2 - 1)$, where $(e_i)_{i\geq 1}$ is a complete orthonormal system in $H = L^2([0, \infty))$ and λ_i is a decreasing sequence of positive numbers such that $\sum_{i=1}^{\infty} \lambda_i^2 < \infty$. Suppose that $E(F^2) = 1$. Then, if $\lambda_N \neq 0$ for some $N > 4$, we obtain

$$\sup_{x\in\mathbb{R}} |p_F(x) - \phi(x)| \leq C_{N,\lambda_N} \sqrt{\sum_{i=1}^{\infty} \lambda_i^4}.$$

We can also establish the uniform convergence of densities in the framework of the Breuer–Major theorem. Fix $q \geq 2$ and consider the sequence

$$V_n = \frac{1}{\sqrt{n}} \sum_{k=1}^{n} \sum_{j=d}^{q} a_j h_j(Y_k), \quad a_d \neq 0,$$

where $Y = (Y_k)_{k\in\mathbb{Z}}$ is a centered Gaussian stationary sequence with unit variance and covariance $\rho(v)$. The following result was proved by Hu *et al.* (2015).

Theorem 8.6.3 *Suppose that the spectral density f_ρ of Y satisfies $\log(f_\rho) \in L^1([-\pi, \pi])$. Assume that $\sum_{v\in\mathbb{Z}} |\rho(v)|^d < \infty$. Set $\sigma^2 := q!a_q^2 \sum_{v\in\mathbb{Z}} \rho(v)^q \in (0, \infty)$. Then, for any $p \geq 1$, there exists an n_0 such that*

$$\sup_{n\geq n_0} E\left(\|DV_n\|_H^{-p}\right) < \infty. \tag{8.29}$$

Therefore, if $q = d$ and $F_n = V_n/\sqrt{E(V_n^2)}$, we have

$$\sup_{x\in\mathbb{R}} |p_{F_n}(x) - \phi(x)| \leq c \sqrt{E(F_n^4) - 3}.$$

Sketch of the proof of Theorem 8.6.3 From the non-causal representation $Y_k = \sum_{j=0}^{\infty} \psi_j w_{k-j}$, where $w = (w_k)_{k\in\mathbb{Z}}$ is a discrete Gaussian white noise, it follows that

$$\|DV_n\|_H^2 \geq \frac{1}{n} \sum_{m=1}^{n} \left(\sum_{k=m}^{n} \sum_{j=d}^{q} a_j h_j'(Y_k)\psi_{k-m}\right)^2 := K_n.$$

Fix N and consider a block decomposition $K_n = \sum_{i=1}^{N} K_n^i$, where K_n^i is the sum of n/N squares. We use the estimate

$$K_n^{-p/2} \leq \prod_{i=1}^{N} (K_n^i)^{-p/(2N)},$$

and apply the Carbery–Wright inequality (see Carbery and Wright, 2001) to control the expectation of $(K_n^i)^{-p/(2N)}$ if $p/(2N)$ is small enough. This inequality says that there is a universal constant $c > 0$ such that, for any polynomial $Q \colon \mathbb{R}^n \to \mathbb{R}$ of degree at most d and any $\alpha > 0$, we have

$$\mathrm{E}(Q(X_1, \ldots, X_n)^2)^{1/(2d)} \mathrm{P}(|Q(X_1, \ldots, X_n)| \leq \alpha) \leq cd\alpha^{1/d},$$

where $X_1, \ldots, X_n$ are independent random variables with law $N(0, 1)$. ☐

This theorem can be applied to the increments of a fractional Brownian motion with Hurst parameter $\mathrm{H} \in (0, 1)$; that is, $Y_k = B_k^{\mathrm{H}} - B_{k-1}^{\mathrm{H}}$, $k \geq 1$. In this case, the spectral density satisfies the required conditions. As a consequence, we obtain the uniform convergence of densities to ϕ for the sequence of Hermite variations $F_n = V_n / \mathrm{E}(V_n^2)$, where

$$V_n = \frac{1}{\sqrt{n}} \sum_{k=1}^{n} h_q(n^{\mathrm{H}} \Delta_{k,n} B^{\mathrm{H}}), \quad q \geq 2,$$

for $0 < \mathrm{H} < 1 - 1/(2q)$, where $\Delta_{k,n} B^{\mathrm{H}} = B_{k/n}^{\mathrm{H}} - B_{(k-1)/n}^{\mathrm{H}}$. In the particular case $q = 2$, we need $\mathrm{H} \in (0, \frac{3}{4})$ and we have

$$\sup_{x \in \mathbb{R}} |p_{F_n}(x) - \phi(x)| \leq c \sqrt{\mathrm{E}(F_n^4) - 3} \leq c_{\mathrm{H}} \begin{cases} n^{-1/2} & \text{if } \mathrm{H} \in (0, \frac{5}{8}), \\ n^{-1/2}(\log n)^{3/2} & \text{if } \mathrm{H} = \frac{5}{8}, \\ n^{4H-3} & \text{if } \mathrm{H} \in (\frac{5}{8}, \frac{3}{4}). \end{cases}$$

The following further results on the uniform convergence of densities were also proved by Hu *et al.* (2014b).

(i) One can show the uniform approximation of the mth derivative of p_F by the corresponding mth derivative of the Gaussian density $\phi^{(m)}$, under the stronger assumption $\mathrm{E}(\|DF\|_H^{-\beta}) < \infty$, for some $\beta > 6m + 6(\lfloor m/2 \rfloor \vee 1)$.

(ii) Consider a d-dimensional vector F, whose components are in a fixed chaos and which is such that $\mathrm{E}((\det \gamma_F)^{-p}) < \infty$ for all p, where γ_F denotes the Malliavin matrix of F. In this case, for any multi-index

$\beta = (\beta_1, \ldots, \beta_k)$, $1 \le \beta_i \le d$, one can show that

$$\sup_{x \in \mathbb{R}^d} \left| \partial_\beta f_F(x) - \partial_\beta \phi_d(x) \right| \le c \left(|C - I_d|^{1/2} + \sum_{j=1}^{d} \sqrt{E(F_j^4) - 3(E(F_j^2))^2} \right)$$

where C is the covariance matrix of F, I_d is the identity matrix of order d, ϕ_d is the standard d-dimensional normal density, and $\partial_\beta = \partial^k / (\partial x_{\beta_1} \cdots \partial x_{\beta_k})$.

8.7 Noncentral Limit Theorems

In this section we present some results on convergence in distribution to a mixture of normal densities, using the techniques of Malliavin calculus.

Definition 8.7.1 Let $(F_n)_{n \ge 1}$ be a sequence of random variables defined on a probability space $(\Omega, \mathcal{F}, P)$. Let F be a random variable defined on some extended probability space $(\Omega', \mathcal{F}', P')$. We say that F_n *converges stably* to F if

$$(F_n, Y) \xrightarrow{\mathcal{L}} (F, Y) \tag{8.30}$$

as $n \to \infty$, for every bounded $\mathcal{F}$-measurable random variable Y.

Suppose that B is a Brownian motion defined on a probability space $(\Omega, \mathcal{F}, P)$ such that $\mathcal{F}$ is generated by B.

Theorem 8.7.2 (Nourdin and Nualart, 2010) *Let $F_n = \delta(u_n)$, where $u_n \in \mathbb{D}^{2,2}(H)$. Suppose that $\sup_n E(|F_n|) < \infty$ and there exists a nonnegative random variable S such that:*

(i) $\langle u_n, DF_n \rangle_H \xrightarrow{L^1} S^2$ *as $n \to \infty$; and,*

(ii) *for all $h \in H$, $\langle u_n, h \rangle_H \xrightarrow{L^1} 0$ as $n \to \infty$.*

Then F_n converges stably to ηS, where η is an $N(0,1)$ random variable independent of B.

Proof It suffices to show that

$$\xi_n := (F_n, B) \xrightarrow{\mathcal{L}} \xi := (F_\infty, B),$$

where F_∞ satisfies, for any $\lambda \in \mathbb{R}$,

$$E(e^{i\lambda F_\infty} | B) = e^{-\lambda^2 S^2 / 2}. \tag{8.31}$$

Since F_n is bounded in $L^1(\Omega)$, the sequence ξ_n is tight. Assume that ξ is the limit in law of a certain subsequence of ξ_n, denoted also by ξ_n. Then

$\xi = (F_\infty, B)$ and it suffices to show that (8.31) holds. Fix $h_1, \ldots, h_n \in H$, let $Y = \varphi(B(h_1), \ldots, B(h_m))$, with $\varphi \in C_b^\infty(\mathbb{R}^m)$, and set

$$\phi_n(\lambda) = \mathrm{E}(e^{i\lambda F_n} Y).$$

We will compute the limit of $\phi_n'(\lambda)$ in two different ways:

1. Using that F_n is bounded in $L^1(\Omega)$ and the convergence in law of ξ_n, we obtain

$$\phi_n'(\lambda) = i\mathrm{E}(e^{i\lambda F_n} F_n Y) \to i\mathrm{E}(e^{i\lambda F_\infty} F_\infty Y). \tag{8.32}$$

2. Using Malliavin calculus we can write

$$\begin{aligned} \phi_n'(\lambda) = i\mathrm{E}(e^{i\lambda F_n} F_n Y) &= i\mathrm{E}(e^{i\lambda F_n} \delta(u_n) Y) \\ &= i\mathrm{E}(\langle D(e^{i\lambda F_n} Y), u_n \rangle_H) \\ &= -\lambda \mathrm{E}(e^{i\lambda F_n} \langle u_n, DF_n \rangle_H Y) + i\mathrm{E}(e^{i\lambda F_n} \langle u_n, DY \rangle_H). \end{aligned}$$

Then, conditions (i) and (ii) of the theorem imply that

$$\phi_n'(\lambda) = i\mathrm{E}(e^{i\lambda F_n} F_n Y) \to i\mathrm{E}(e^{i\lambda F_\infty} S^2 Y). \tag{8.33}$$

As a consequence, from (8.32) and (8.33) we get

$$i\mathrm{E}(e^{i\lambda F_\infty} F_\infty Y) = -\lambda \mathrm{E}(e^{i\lambda F_\infty} S^2 Y).$$

This leads to a linear differential equation satisfied by the conditional characteristic function of F_∞,

$$\frac{\partial}{\partial \lambda} \mathrm{E}(e^{i\lambda F_\infty} | B) = -S^2 \lambda \mathrm{E}(e^{i\lambda F_\infty} | B),$$

and we obtain (8.31), which completes the proof. $\qquad\square$

It turns out that

$$\langle u_n, DF_n \rangle_H = \|u_n\|_H^2 + \langle u_n, \delta(Du_n) \rangle_H.$$

Therefore, a sufficient condition for (i) to hold is:

(i') $\|u_n\|_H^2 \xrightarrow{L^1} S^2$ and $\langle u_n, \delta(Du_n) \rangle_H \xrightarrow{L^1} 0$.

This result should be compared with the *asymptotic Ray–Knight theorem* for Brownian martingales (see Revuz and Yor, 1999, Theorem (2.3), p. 524):

Proposition 8.7.3 *If $u_n = (u_n(t))_{t \in [0,T]}$, $n \geq 1$, is a sequence of square integrable adapted processes then $F_n = \delta(u_n) = \int_0^T u_n(s) dB_s$ and the stable convergence of F_n to $N(0, S^2)$ is implied by the following conditions:*

(i) $\int_0^t u_n(s)ds \xrightarrow{P} 0$, *uniformly in t;*

(ii) $\int_0^T u_n(s)^2 ds \rightarrow S^2$ *in* $L^1(\Omega)$.

Theorem 8.7.2 can be extended to multiple divergences. The proof is similar and is therefore omitted.

Theorem 8.7.4 (Nourdin and Nualart, 2010) *Fix an integer $q \geq 1$. Let $F_n = \delta^q(u_n)$, where u_n is a symmetric element in $\mathbb{D}^{2q,2q}(H^{\otimes q})$. Suppose that $\sup_n \|F_n\|_{q,p} < \infty$ for any $p \geq 2$ and*

(i) $\langle u_n, D^q F_n \rangle_{H^{\otimes q}} \xrightarrow{L^1} S^2$,

(ii) $\langle u_n, h \rangle_{H^{\otimes q}} \rightarrow 0$ *in* L^1 *for any* $h \in H^{\otimes q}$, *and* $u_n \otimes_j D^j F_n \rightarrow 0$ *in* L^2 *for all* $j = 1, \ldots, q-1$.

Then F_n converges stably to ηS, where η is an $N(0,1)$ random variable independent of B.

Stein's method does not seem to work in the case of a random variance. Nevertheless, it is possible to derive rates of convergence using different approaches. The next result provides a rate of convergence in the context of a mixture of normal laws, and its proof is based on the interpolation method.

Theorem 8.7.5 (Nourdin *et al.*, 2016a) *Let $F = \delta(u)$, where $u \in \mathbb{D}^{2,2}(H)$. Let $S \geq 0$ be a random variable such that $S^2 \in \mathbb{D}^{1,2}$, and let η be an $N(0,1)$ random variable independent of B. Then, for any $\varphi \in C_b^3(\mathbb{R})$,*

$$|\mathrm{E}(\varphi(F)) - \mathrm{E}(\varphi(S\eta))| \leq \tfrac{1}{2}\|\varphi''\|_\infty \mathrm{E}(|\langle u, DF \rangle_H - S^2|)$$
$$+ \tfrac{1}{3}\|\varphi'''\|_\infty \mathrm{E}(|\langle u, DS^2 \rangle_H|).$$

Proof Notice that S is not necessarily in $\mathbb{D}^{1,2}$. However, if we fix $\epsilon > 0$ and set $S_\epsilon = \sqrt{S^2 + \epsilon}$ then $S_\epsilon \in \mathbb{D}^{1,2}$. Let $g(t) = \mathrm{E}(\varphi(\sqrt{t}F + \sqrt{1-t}S_\epsilon\eta))$, $t \in [0,1]$. Then

$$\mathrm{E}(\varphi(F)) - \mathrm{E}(\varphi(S_\epsilon\eta)) = g(1) - g(0) = \int_0^1 g'(t)dt.$$

Integrating by parts using $F = \delta(u)$ yields

$$g'(t) = \tfrac{1}{2}\mathrm{E}\left(\varphi'(\sqrt{t}F + \sqrt{1-t}S_\epsilon\eta)\left(\frac{F}{\sqrt{t}} - \frac{S_\epsilon\eta}{\sqrt{1-t}}\right)\right)$$

$$= \tfrac{1}{2}\mathrm{E}\left(\varphi''(\sqrt{t}F + \sqrt{1-t}S_\epsilon\eta)\right.$$

$$\left. \times \left(\langle u, DF \rangle_H + \frac{\sqrt{1-t}}{\sqrt{t}}\eta\langle u, DS_\epsilon \rangle_H - S_\epsilon^2\right)\right).$$

Integrating again by parts with respect to the law of η yields

$$g'(t) = \tfrac{1}{2} \mathrm{E}\left(\varphi''(\sqrt{t}F + \sqrt{1-t}S_\epsilon \eta)\left(\langle u, DF \rangle_{\mathrm{H}} - S_\epsilon^2\right)\right)$$
$$+ \frac{1-t}{4\sqrt{t}} \mathrm{E}\left(\varphi'''(\sqrt{t}F + \sqrt{1-t}S_\epsilon \eta)\langle u, DS^2 \rangle_H\right),$$

where we have used the fact that $S_\epsilon DS_\epsilon = \tfrac{1}{2}DS_\epsilon^2 = \tfrac{1}{2}DS^2$. Finally, integrating over t yields

$$|\mathrm{E}(\varphi(F)) - \mathrm{E}(\varphi(S_\epsilon \eta))| \le \tfrac{1}{2} \|\varphi''\|_\infty \mathrm{E}(|\langle u, DF \rangle_H - S^2 - \epsilon|)$$
$$+ \|\varphi'''\|_\infty \mathrm{E}(|\langle u, DS^2 \rangle_H|) \int_0^1 \frac{1-t}{4\sqrt{t}} dt,$$

and the conclusion follows because

$$\int_0^1 \frac{1-t}{4\sqrt{t}} dt = \frac{1}{3}.$$

$\square$

Noncentral limit theorems arise in a wide range of asymptotic problems, and the techniques of Malliavin calculus have proved to be useful to show the convergence in law to a mixture of Gaussian laws and to analyze the rate of convergence. For instance, we mention the following:

1. the asymptotic behavior of weighted Hermite variations of a fractional Brownian motion (see Nourdin *et al.*, 2010a; Nourdin and Nualart, 2010; Nourdin *et al.*, 2016b);
2. the approximation by Riemann sums of symmetric integrals with respect to fractional Brownian motions for critical values of the Hurst parameter and Itô's formulas in law (see Burdzy and Swanson, 2010; Nourdin *et al.*, 2010b; Harnett and Nualart, 2012, 2013, 2015; Binotto *et al.*, 2018);
3. the fluctuation of the error in numerical approximation for stochastic differential equations (see Hu *et al.*, 2016).

Exercises

8.1   Show Proposition 8.1.7.

8.2 Let X_1 and X_2 two random variables with laws $N(0, \sigma_1^2)$ and $N(0, \sigma_2^2)$, respectively, where $\sigma_1 > 0$, $\sigma_2 > 0$, and $\sigma_1 \geq \sigma_2$. Show that

$$d_{\mathrm{TV}}(X_1, X_2) \leq \frac{2}{\sigma_1^2} |\sigma_1^2 - \sigma_2^2|,$$

$$d_{\mathrm{Kol}}(X_1, X_2) \leq \frac{1}{\sigma_1^2} |\sigma_1^2 - \sigma_2^2|,$$

$$d_{\mathrm{W}}(X_1, X_2) \leq \frac{\sqrt{2/\pi}}{\sigma_1^2} |\sigma_1^2 - \sigma_2^2|.$$

8.3 Show that the sequence $F^{(n)} = \int_0^1 u_t^{(n)} dB_t$ where B is a Brownian motion and

$$u_t^{(n)} = \sqrt{2} n t^n \exp(B_t(1-t)), \quad 0 \leq t \leq 1,$$

converges in law to the distribution $N(0, 1)$.

8.4 Let $B = (B_t)_{t \geq 0}$ be a standard Brownian motion. Consider the sequence of Itô integrals

$$F_n = \sqrt{n} \int_0^1 t^n B_t dB_t, \quad n \geq 1.$$

Show that the sequence F_n converges stably to ηS as $n \to \infty$, where η is a random variable independent of B with law $N(0, 1)$ and $S = |B_1| / \sqrt{2}$.

8.5 Let $B^{\mathrm{H}} = (B_t^{\mathrm{H}})_{t \geq 0}$ be a fractional Brownian motion with Hurst parameter $H > \frac{1}{2}$. Consider the sequence of random variables $F_n = \delta(u_n), n \geq 1$, where

$$u_n(t) = n^H t^n B_t 1_{[0,1]}(t).$$

Show that, as $n \to \infty$, the sequence F_n converges stably to ηS, where η is a random variable that is independent of B with law $N(0, 1)$ and $S = c_{\mathrm{H}} |B_1^{\mathrm{H}}|$, with $c_{\mathrm{H}} = \sqrt{H(2H - 1)\Gamma(2H - 1)}$.

8.6 In Exercises 8.4 and 8.5 find the rates of convergence to zero of $|E(\varphi(F_n)) - E(\varphi(S\eta))|$, where $\varphi \in C_b^3(\mathbb{R})$.

9

Jump Processes

In this chapter we introduce Lévy processes and Poisson random measures. We construct a stochastic calculus, with respect to a Poisson random measure associated with a Lévy process, which includes Itô's formula, the integral representation theorem, and Girsanov's theorem. Finally, we define multiple stochastic integrals and prove the Wiener chaos decomposition for general Poisson random measures. For more detailed accounts of these topics we refer to Sato (1999) and Appelbaum (2009).

9.1 Lévy Processes

A stochastic process $X = (X_t)_{t \geq 0}$ is said to have càdlàg paths if almost all its sample paths are càdlàg, that is, continuous from the right with limits from the left.

Definition 9.1.1 A real-valued process $L = (L_t)_{t \geq 0}$ defined on a probability space $(\Omega, \mathcal{F}, P)$ is called a *Lévy process* if it satisfies the following conditions.

 (i) Almost surely $L_0 = 0$.
 (ii) For all $0 \leq t_1 < \cdots < t_n$ the increments $L_{t_n} - L_{t_{n-1}}, \ldots, L_{t_2} - L_{t_1}$ are independent random variables.
 (iii) L has stationary increments; that is, if $s < t$, the increment $L_t - L_s$ has the same law as L_{t-s}.
 (iv) L is continuous in probability.
 (v) L has càdlàg paths.

Properties (ii) and (iii) imply that for any $t > 0$ the random variable L_t is infinitely indivisible. Moreover, the characteristic function of L_t has the following Lévy–Khintchine representation:

$$E(\exp(iuL_t)) = \exp(t\psi(u)), \quad \text{for all } u \in \mathbb{R}, \tag{9.1}$$

where ψ is the characteristic exponent of the random variable L_1 and is given by

$$\psi(u) = i\beta u - \frac{\alpha^2 u^2}{2} + \int_{\mathbb{R}} (e^{iuz} - 1 - iuz\mathbf{1}_{\{|z|\leq 1\}})\nu(dz). \qquad (9.2)$$

Here $\beta \in \mathbb{R}$, $\alpha^2 \geq 0$, and ν is the Lévy measure of L, which is a σ-finite measure on $\mathcal{B}(\mathbb{R})$ satisfying $\nu(\{0\}) = 0$ and

$$\int_{\mathbb{R}} (1 \wedge |z|^2)\nu(dz) < \infty.$$

We call (β, α^2, ν) the characteristic triplet of L. Moreover, β is called the drift term and α the diffusion coefficient.

Conversely, it can be proved that given a triplet (β, α^2, ν), with $\beta \in \mathbb{R}$, $\alpha^2 \geq 0$, and ν a Lévy measure, there exists a unique (in law) Lévy process L such that (9.1) and (9.2) hold (see Sato, 1999, Theorems 7.10 and 8.1).

Example 9.1.2 (Brownian motion with drift) The process $X_t = \alpha B_t + \beta t$ is a Lévy process with continuous trajectories and characteristic triplet $(\beta, \alpha^2, 0)$. Moreover, any Lévy process with continuous trajectories is a Brownian motion with drift.

Example 9.1.3 (Poisson process) Let $(\tau_i)_{i\geq 1}$ be a sequence of independent exponential random variables with parameter $\lambda > 0$, and let $T_n = \sum_{i=1}^{n} \tau_i$. Then, the process $N = (N_t)_{t\geq 0}$ defined by

$$N_t = \sum_{n\geq 1} \mathbf{1}_{\{t\geq T_n\}}$$

is called a Poisson process with intensity λ. For any $t > 0$, the random variable N_t follows a Poisson distribution with parameter λt (see Exercise 9.1). The Poisson process is a Lévy process with characteristic triplet $(0, 0, \nu)$, where $\nu = \lambda \delta_1$ and δ_1 is the Dirac measure supported on $\{1\}$. Observe that the trajectories of N are piecewise constant with jumps of size 1.

Example 9.1.4 (Compound Poisson process) The compound Poisson process with jump intensity $\lambda > 0$ and jump size distribution μ is a stochastic process $(X_t)_{t\geq 0}$ defined by

$$X_t = \sum_{i=1}^{N_t} Y_i,$$

where $(Y_i)_{i\geq 1}$ is a sequence of independent random variables with law μ and N is a Poisson process with intensity λ and independent of $(Y_i)_{i\geq 1}$. By convention, $X_t = 0$ if $N_t = 0$. In other words, a compound Poisson process is a piecewise constant process which jumps at the jump times of a Poisson process and whose jump sizes are i.i.d. random variables with a given law. This defines a Lévy process with characteristic triplet $(0, 0, \nu)$, where $\nu = \lambda\mu$.

9.2 Poisson Random Measures

Let $(Z, \mathcal{Z}, m)$ be a measure space, where Z is a complete separable metric space, $\mathcal{Z}$ is the Borel σ-field of Z, and m is a σ-finite atomless measure. Consider the class $\mathcal{Z}_m = \{A \in \mathcal{Z} : m(A) < \infty\}$.

A Poisson random measure is defined as the following integer-valued random measure.

Definition 9.2.1   A *Poisson random measure* with intensity m is a collection of random variables $M = \{M(A), A \in \mathcal{Z}_m\}$, defined on some probability space $(\Omega, \mathcal{F}, P)$, such that:

(i) for every $A \in \mathcal{Z}_m$, $M(A)$ has a Poisson distribution with parameter $m(A)$;
(ii) for every $A_1, \ldots, A_n \in \mathcal{Z}_m$, pairwise disjoint, $M(A_1), \ldots, M(A_n)$ are independent.

We then define the compensated Poisson random measure as

$$\hat{M} = \{\hat{M}(A) = M(A) - m(A), A \in \mathcal{Z}_m\}.$$

The following result concerns the construction of Poisson random measures in such a way that they are point measures on a finite or countable number of different points.

Theorem 9.2.2   *Let m be a σ-finite atomless measure on a complete separable metric space $(Z, \mathcal{Z}, m)$. Then there exists a Poisson random measure M with intensity m. Moreover, M has the form*

$$M(A) = \sum_{j=1}^{N} \delta_{X_j}(A), \quad A \in \mathcal{Z}_m, \tag{9.3}$$

where X_j, $j = 1, \ldots, N$, are random points, $N \le \infty$ is a random variable, and almost surely $X_j \ne X_k$ for $j \ne k$.

Proof   First assume that $m(Z) < \infty$. In this case, we can consider a sequence of independent random variables $N, X_1, X_2, \ldots$, where N has a Poisson distribution with intensity $m(Z)$ and $P(X_j \in A) = m(A)/m(Z)$, for all

$j \geq 1$ and $A \in \mathcal{Z}_m$. For all $A \in \mathcal{Z}_m$, we set

$$M(A) = \sum_{j=1}^{N} \mathbf{1}_A(X_j).$$

Clearly, M satisfies (9.3) and has exactly N atoms because m is atomless. We next compute the finite-dimensional distributions of M.

By conditioning first on N, we find that, for every collection of pairwise disjoint sets $A_1, \ldots, A_k \in \mathcal{Z}_m$ and $\xi_1, \ldots, \xi_k \in \mathbb{R}$,

$$\mathrm{E}\left(\exp\left(i \sum_{j=1}^{k} \xi_j M(A_j) \right) \right) = \mathrm{E}\left(\prod_{\ell=1}^{N} \exp\left(i \sum_{j=1}^{k} \xi_j \mathbf{1}_{A_j}(X_\ell) \right) \right)$$

$$= \mathrm{E}\left(\left(\mathrm{E}\left(\exp\left(i \sum_{j=1}^{k} \xi_j \mathbf{1}_{A_j}(X_1) \right) \right) \right)^{N} \right).$$

Because the A_j are pairwise disjoint, the indicator function of $(A_1 \cup \cdots \cup A_k)^c$ is equal to $1 - \sum_{j=1}^{k} \mathbf{1}_{A_j}$ and hence, for all $z \in Z$,

$$\exp\left(i \sum_{j=1}^{k} \xi_j \mathbf{1}_{A_j}(z) \right) = \sum_{j=1}^{k} \mathbf{1}_{A_j}(z) e^{i\xi_j} + 1 - \sum_{j=1}^{k} \mathbf{1}_{A_j}(z)$$

$$= 1 + \sum_{j=1}^{k} \mathbf{1}_{A_j}(z)(e^{i\xi_j} - 1).$$

Therefore

$$\mathrm{E}\left(\exp\left(i \sum_{j=1}^{k} \xi_j \mathbf{1}_{A_j}(X_1) \right) \right) = 1 + \sum_{j=1}^{k} \left(m(A_j)/m(Z) \right) (e^{i\xi_j} - 1),$$

and hence

$$\mathrm{E}\left(\exp\left(i \sum_{j=1}^{k} \xi_j M(A_j) \right) \right) = \mathrm{E}\left(\left(1 + \sum_{j=1}^{k} \left(m(A_j)/m(Z) \right) (e^{i\xi_j} - 1) \right)^{N} \right).$$

Finally, using the fact that $\mathrm{E}(r^N) = \exp(-m(Z)(1 - r))$, for all $r \in \mathbb{R}$, we conclude that

$$\mathrm{E}\left(\exp\left(i \sum_{j=1}^{k} \xi_j M(A_j) \right) \right) = \exp\left(-\sum_{j=1}^{k} m(A_j)(1 - e^{i\xi_j}) \right).$$

This implies that the random variables $M(A_j)$, $1 \leq j \leq k$, are independent and have Poisson distributions with parameter $m(A_j)$. So, the desired result follows in the case where $m(Z) < \infty$. In the general case, we can find pairwise disjoint sets $Z_1, Z_2, \ldots \in \mathcal{Z}$ such that $Z = \cup_{j=1}^{\infty} Z_j$ and $m(Z_j) < \infty$ for all $j \geq 1$. We can then construct independent Poisson random measures $M_1, M_2, \ldots$, as in the preceding construction, where each M_j is based only on subsets of Z_j. Then, we set $M(A) = \sum_{j=1}^{\infty} M_j(A \cap Z_j)$ for all $A \in \mathcal{Z}_m$. Because a sum of independent Poisson random variables has a Poisson distribution, it follows that M is a Poisson random measure with intensity m. Moreover, M has the form (9.3) with $N = \infty$. This concludes the proof.

$\square$

Definition 9.2.3 The canonical probability space $(\Omega, \mathcal{F}, P)$ of a Poisson random measure is called the *Poisson space* and is defined as follows:

- Ω is the set of nonnegative integer-valued measures on Z, that is,

$$\Omega = \left\{ \omega = \sum_{j=0}^{n} \delta_{z_j}, n \in \mathbb{N} \cup \{\infty\}, z_j \in Z \right\}; \qquad (9.4)$$

- P is the unique probability measure such that the canonical mapping

$$M(A)(\omega) = \omega(A), \quad A \in \mathcal{Z}_m,$$

is a Poisson random measure;
- $\mathcal{F}$ is the P-completion of the σ-field generated by the canonical mapping.

By taking the law of the Poisson random measure constructed in Theorem 9.2.2, we deduce that the Poisson space exists and that

$$P(\omega \in \Omega : \exists z \in Z : \omega(\{z\}) > 1) = 0. \qquad (9.5)$$

In the next chapter the Poisson space will play an important role in the construction of the Malliavin calculus with respect to a Poisson random measure. For the rest of this chapter, we assume that $(\Omega, \mathcal{F}, P)$ is an arbitrary probability space.

Example 9.2.4 (Jump measure of a Lévy process) Let $L = (L_t)_{t \geq 0}$ be a Lévy process with Lévy measure ν. Set $\mathbb{R}_0 = \mathbb{R} \setminus \{0\}$ and consider the set

$$\mathcal{A}_\nu = \{A \in \mathcal{B}(\mathbb{R}_0) : \nu(A) < \infty\}.$$

For any $0 \leq s \leq t$ and $A \in \mathcal{A}_\nu$, define

$$N([s, t], A) = \sum_{s \leq r \leq t} \mathbf{1}_A(\Delta L_r), \qquad (9.6)$$

where $\Delta L_t = L_t - L_{t-}$. By convention, $L_{0-} = 0$.

The random variable N counts the number of jumps of the Lévy process

L between s and t such that their sizes fall into A. One can show (see Sato, 1999, Theorem 19.2) that N can be extended to a Poisson random measure in $(\mathbb{R}_+ \times \mathbb{R}_0, \mathcal{B}(\mathbb{R}_+ \times \mathbb{R}_0))$ with intensity $dt\, \nu(dz)$ and, for all $A \in \mathcal{A}_\nu$, $\nu(A) = E(N([0, 1], A))$. We write $N(t, A)$ for $N([0, t], A)$. The measure $\hat{N}(t, A) = N(t, A) - \nu(A)t$ is the corresponding compensated Poisson random measure.

For any $t \geq 0$, we denote by $\mathcal{F}_t$ the σ-field generated by the random variables $\{\hat{N}(s, A), A \in \mathcal{A}_\nu, s \leq t\}$ and the P-null sets of $\mathcal{F}$. We call $(\mathcal{F}_t)_{t\geq 0}$ the natural filtration of N. Then, for any $A \in \mathcal{A}_\nu$, the process $(\hat{N}(t, A))_{t\geq 0}$ is an $\mathcal{F}_t$-martingale (see Exercise 9.2).

9.3 Integral with respect to a Poisson Random Measure

Consider a Poisson random measure M on a complete separable metric space $(Z, \mathcal{Z}, m)$, where m is a σ-finite atomless measure. In this section we define the integral of functions in $L^2(Z)$ with respect to the compensated Poisson random measure $\hat{M}$.

We consider the set $\mathcal{S}$ of simple functions

$$h(z) = \sum_{j=1}^{n} a_j \mathbf{1}_{A_j}(z), \tag{9.7}$$

where $n \geq 1$ is an integer, $a_1, \ldots, a_n \in \mathbb{R}$, and $A_1, \ldots, A_n \in \mathcal{Z}_m$. We define the integral of a simple function $h \in \mathcal{S}$ of the form (9.7), with respect to $\hat{M}$, by

$$\int_Z h(z)\hat{M}(dz) = \sum_{j=1}^{n} a_j \hat{M}(A_j),$$

and we denote it by $\hat{M}(h)$. This defines a linear mapping $h \to \hat{M}(h)$ from $\mathcal{S} \subset L^2(Z)$ into $L^2(\Omega)$ with the following properties:

$$E(\hat{M}(h)) = 0, \quad E(\hat{M}(h)^2) = \|h\|_{L^2(Z)}^2 = \int_Z h^2(z)m(dz).$$

As $\mathcal{S}$ is dense in $L^2(Z)$, we can extend the definition of the integral as a linear isometry from $L^2(Z)$ to $L^2(\Omega)$. In particular, the following isometric relation holds for every $h, g \in L^2(Z)$:

$$E(\hat{M}(h)\hat{M}(g)) = \int_Z h(z)g(z)m(dz).$$

Moreover, for every $h \in L^2(Z)$, the random variable $\hat{M}(h)$ has an infinitely divisible law, with Lévy–Khinchine characteristic exponent

$$\psi(u) = \log \mathrm{E}\left(e^{iu\hat{M}(h)}\right) = \int_Z (e^{iuh(z)} - 1 - iuh(z))m(dz), \quad \text{for all } u \in \mathbb{R}.$$

Example 9.3.1 Consider the jump measure N associated with a Lévy process L with characteristic triplet (β, α^2, ν). Since $\int_{\{|z| \leq 1\}} z^2 \nu(dz) < \infty$, we can consider the integral

$$\int_0^t \int_{\{|z| \leq 1\}} z \, \hat{N}(ds, dz).$$

Then we have the following representation of the Lévy process (see Sato, 1999, Chapter 4, and Appelbaum, 2009, Theorem 2.4.16):

$$L_t = \beta t + \alpha B_t + \int_0^t \int_{\{|z| \leq 1\}} z \, \hat{N}(ds, dz) + \int_0^t \int_{\{|z| > 1\}} z \, N(ds, dz), \quad (9.8)$$

where B is a Brownian motion independent of N and

$$\int_0^t \int_{\{|z| > 1\}} z \, N(ds, dz) = \sum_{0 \leq r \leq t} \Delta L_r \mathbf{1}_{\{|z| > 1\}}(\Delta L_r),$$

which is well defined since $\int_{\{|z| > 1\}} \nu(dz) < \infty$.

9.4 Stochastic Integrals with respect to the Jump Measure of a Lévy Process

In this section we consider the Poisson random measure N associated with a Lévy process, and we define a stochastic integral with respect to the compensated Poisson random measure $\hat{N}$ for the class of predictable and square integrable processes defined below. We recall that $(\mathcal{F}_t)_{t \geq 0}$ denotes the natural filtration of N.

A stochastic process $u = \{u(t, z), t \geq 0, z \in \mathbb{R}_0\}$ is called adapted if $u(t, z)$ is $\mathcal{F}_t$-measurable for all $t \geq 0$ and $z \in \mathbb{R}_0$. Moreover, u is called predictable if it is measurable with respect to the σ-field generated by the sets

$$\{B \times (s, t] \times A, B \in \mathcal{F}_s, 0 \leq s < t, A \in \mathcal{B}(\mathbb{R}_0)\}.$$

Notice that any adapted and left-continuous process (in t) is predictable.

We denote by $L^2(\mathcal{P})$ the set of stochastic processes $u \colon \Omega \times \mathbb{R}_+ \times \mathbb{R}_0 \to \mathbb{R}$ that are predictable and satisfy

$$\int_{\mathbb{R}_+} \int_{\mathbb{R}_0} \mathrm{E}(u^2(t, z))\nu(dz)dt < \infty.$$

Let $\mathcal{E}$ denote the set of elementary and predictable processes of the form

$$u(t, z) = \sum_{j=1}^{m} \sum_{i=0}^{n-1} F_{i,j} \mathbf{1}_{(t_i, t_{i+1}]}(t) \mathbf{1}_{A_j}(z), \tag{9.9}$$

where $0 \le t_0 < \cdots < t_n$, all $F_{i,j}$ belong to $\mathcal{F}_{t_i}$ and are bounded, and $A_1, \ldots, A_m$ are pairwise disjoint subsets of $\mathcal{A}_\nu$.

We define the integral of $u \in \mathcal{E}$ of the form (9.9) with respect to $\hat{N}$ by

$$\int_{\mathbb{R}_+} \int_{\mathbb{R}_0} u(t, z) \hat{N}(dt, dz) = \sum_{j=1}^{m} \sum_{i=0}^{n-1} F_{i,j} \hat{N}((t_i, t_{i+1}], A_j). \tag{9.10}$$

This defines a linear functional with the following properties:

Lemma 9.4.1 *For any $u \in \mathcal{E}$,*

$$\mathrm{E}\left(\int_{\mathbb{R}_+} \int_{\mathbb{R}_0} u(t, z) \hat{N}(dt, dz) \right) = 0 \tag{9.11}$$

and

$$\mathrm{E}\left(\int_{\mathbb{R}_+} \int_{\mathbb{R}_0} u(t, z) \hat{N}(dt, dz) \right)^2 = \int_{\mathbb{R}_+} \int_{\mathbb{R}_0} \mathrm{E}(u^2(t, z)) \nu(dz) dt. \tag{9.12}$$

Proof By the definition of $\hat{N}$, for each $j = 1, \ldots, m$ and $i = 0, \ldots, n-1$,

$$\mathrm{E}(\hat{N}((t_i, t_{i+1}], A_j)) = 0.$$

Hence, using the linearity of the expectation and the fact that $\hat{N}((t_i, t_{i+1}], A_j)$ is independent of $\mathcal{F}_{t_i}$, we obtain (9.11).

In order to prove (9.12) we use again the linearity of the expectation, to write

$$\mathrm{E}\left(\int_{\mathbb{R}_+} \int_{\mathbb{R}_0} u(t, z) \hat{N}(dt, dz) \right)^2$$
$$= \sum_{j,\ell=1}^{m} \sum_{i,p=0}^{n-1} \mathrm{E}\left(F_{i,j} \hat{N}((t_i, t_{i+1}], A_j) F_{p,\ell} \hat{N}((t_p, t_{p+1}], A_\ell) \right).$$

Observe that, for any indices i, j, ℓ, and $p < i$, since $\hat{N}((t_i, t_{i+1}], A_j)$ is independent of $\mathcal{F}_{t_i}$ we have

$$\mathrm{E}\left(F_{i,j} \hat{N}((t_i, t_{i+1}], A_j) F_{p,\ell} \hat{N}((t_p, t_{p+1}], A_\ell) \right)$$
$$= \mathrm{E}\left(F_{i,j} F_{p,\ell} \hat{N}((t_p, t_{p+1}], A_\ell) \right) \mathrm{E}\left(\hat{N}((t_i, t_{i+1}], A_j) \right) = 0,$$

and similarly for $p > i$. Therefore,

$$E\left(\int_{\mathbb{R}_+}\int_{\mathbb{R}_0} u(t,z)\hat{N}(dt,dz)\right)^2$$

$$= \sum_{j,\ell=1}^{m}\sum_{i=0}^{n-1} E\left(F_{i,j}\hat{N}((t_i,t_{i+1}],A_j)F_{i,\ell}\hat{N}((t_i,t_{i+1}],A_\ell)\right).$$

Then, using property (ii) of a Poisson random measure (see Definition 9.2.1), we get

$$E\left(\int_{\mathbb{R}_+}\int_{\mathbb{R}_0} u(t,z)\hat{N}(dt,dz)\right)^2 = \sum_{j=1}^{m}\sum_{i=0}^{n-1} E(F_{i,j}^2)E\left(\hat{N}((t_i,t_{i+1}],A_j)^2\right)$$

$$= \sum_{j=1}^{m}\sum_{i=0}^{n-1} E(F_{i,j}^2)(t_{i+1}-t_i)\nu(A_j),$$

which concludes the proof. $\qquad\square$

The set $\mathcal{E}$ is dense in $L^2(\mathcal{P})$ (see Appelbaum, 2009, Lemma 4.1.1). Therefore, using the isometry property (9.12), we can extend the stochastic integral to the class $L^2(\mathcal{P})$ in such a way that it is a linear isometry from $L^2(\mathcal{P})$ into $L^2(\Omega)$ satisfying (9.11) and (9.12).

We denote by $L_\infty^2(\mathcal{P})$ the set of stochastic processes $u: \Omega \times \mathbb{R}_+ \times \mathbb{R}_0 \to \mathbb{R}$ that are predictable and satisfy

$$\int_0^t\int_{\mathbb{R}_0} E(u^2(s,z))\nu(dz)ds < \infty,$$

for all $t > 0$. For any process $u \in L_\infty^2(\mathcal{P})$, by considering the restriction of u to each interval $[0,t]$, we can define the indefinite integral process

$$\left\{\int_0^t\int_{\mathbb{R}_0} u(s,z)\hat{N}(ds,dz), t \geq 0\right\}.$$

This process is an $\mathcal{F}_t$-martingale (see Appelbaum, 2009, Theorem 4.2.3).

We next extend the stochastic integral to the set $L_{loc}^2(\mathcal{P})$ of stochastic processes $u: \Omega \times \mathbb{R}_+ \times \mathbb{R}_0 \to \mathbb{R}$ that are predictable and satisfy

$$P\left(\int_0^t\int_{\mathbb{R}_0} u^2(s,z)\nu(dz)ds < \infty\right) = 1,$$

for all $t > 0$. The extension of the stochastic integral is based on the following result.

Proposition 9.4.2 *Let* $u \in \mathcal{E}$. *Then, for all* $T, K, \delta > 0$, *we have*

$$P\left(\left|\int_0^T \int_{\mathbb{R}_0} u(t,z)\hat{N}(dt,dz)\right| \geq K\right) \leq \frac{\delta}{K^2} + P\left(\int_0^T \int_{\mathbb{R}_0} u^2(t,z)\nu(dz)dt \geq \delta\right).$$

Proof Let $u \in \mathcal{E}$ be of the form (9.9) with $0 \leq t_0 < \cdots < t_n = T$. Fix $\delta > 0$ and let n_δ be the largest integer less than or equal to n for which

$$\sum_{j=1}^m \sum_{i=0}^{n_\delta-1} F_{i,j}^2(t_{i+1} - t_i)\nu(A_j) \leq \delta.$$

Consider the process

$$u^\delta(t,z) = \sum_{j=1}^m \sum_{i=0}^{n-1} F_{i,j}\mathbf{1}_{\{i \leq n_\delta-1\}}\mathbf{1}_{(t_i,t_{i+1}]}(t)\mathbf{1}_{A_j}(z).$$

Observe that the event $\{i \leq n_{\delta-1}\}$ is $\mathcal{F}_{t_i}$-measurable, so the process u^δ is predictable and elementary. Then, using Chebyshev's inequality, we obtain

$$P\left(\left|\int_0^T \int_{\mathbb{R}_0} u(t,z)\hat{N}(dt,dz)\right| \geq K\right)$$

$$= P\left(\left|\int_0^T \int_{\mathbb{R}_0} u(t,z)\hat{N}(dt,dz)\right| \geq K, n_\delta = n\right)$$

$$+ P\left(\left|\int_0^T \int_{\mathbb{R}_0} u(t,z)\hat{N}(dt,dz)\right| \geq K, n_\delta \neq n\right)$$

$$\leq P\left(\left|\int_0^T \int_{\mathbb{R}_0} u^\delta(t,z)\hat{N}(dt,dz)\right| \geq K\right) + P(n_\delta \neq n)$$

$$\leq \frac{E\left(\int_0^T \int_{\mathbb{R}_0} u^\delta(t,z)^2\nu(dz)dt\right)}{K^2} + P\left(\int_0^T \int_{\mathbb{R}_0} u^2(t,z)\nu(dz)dt \geq \delta\right)$$

$$\leq \frac{\delta}{K^2} + P\left(\int_0^T \int_{\mathbb{R}_0} u^2(t,z)\nu(dz)dt \geq \delta\right).$$

$\square$

Now let $u \in L^2_{loc}(\mathcal{P})$. Then we can find a sequence of processes $u^{(n)}$ in $\mathcal{E}$ (see Exercise 9.3) such that, for all $T > 0$, the sequence

$$\int_0^T \int_{\mathbb{R}_0} |u(t,x) - u^{(n)}(t,z)|^2\nu(dz)dt$$

converges to zero in probability. By Proposition 9.4.2, for any $m, n \in \mathbb{N}$,

$K, \delta > 0$,

$$P\left(\left|\int_0^T \int_{\mathbb{R}_0} \left(u^{(m)}(t,x) - u^{(n)}(t,z)\right) \hat{N}(dt,dz)\right| \geq K\right)$$

$$\leq \frac{\delta}{K^2} + P\left(\int_0^T \int_{\mathbb{R}_0} |u^{(m)}(t,x) - u^{(n)}(t,z)|^2 \nu(dz)dt \geq \delta\right).$$

Therefore, the sequence

$$\int_0^T \int_{\mathbb{R}_0} u^{(n)}(t,z)\hat{N}(dt,dz)$$

is Cauchy in probability, and thus converges in probability. We denote the limit by

$$\int_0^T \int_{\mathbb{R}_0} u(t,z)\hat{N}(dt,dz)$$

and call it the (extended) stochastic integral. The process $(M_t)_{t\geq 0}$ defined by $M_t = \int_0^t \int_{\mathbb{R}_0} u(s,z)\hat{N}(ds,dz)$ is a local martingale (see Appelbaum, 2009, Theorem 4.2.12).

As a consequence of Proposition 9.4.2, we have the following result.

Corollary 9.4.3   *Let $u \in L^2_{loc}(\mathcal{P})$. Then*

$$\lim_{n\to\infty} \int_0^T \int_{\{|z|\geq 1/n\}} u(s,z)\hat{N}(ds,dz) = \int_0^T \int_{\mathbb{R}_0} u(s,z)\hat{N}(ds,dz),$$

in probability.

Proof   By Exercise 9.4, Proposition 9.4.2 extends to all $u \in L^2_{loc}(\mathcal{P})$. Then, for all $\delta, \epsilon > 0$, we have

$$P\left(\left|\int_0^T \int_{\{0<|z|<1/n\}} u(t,z)\hat{N}(dt,dz)\right| \geq \epsilon\right)$$

$$\leq \frac{\delta}{\epsilon^2} + P\left(\int_0^T \int_{\{0<|z|<1/n\}} u^2(t,z)\nu(dz)dt \geq \delta\right),$$

from which the desired result follows. $\square$

9.5 Itô's Formula

We next give Itô's formula for jump processes. We assume that $L = (L_t)_{t\geq 0}$ is a Lévy process with the representation (9.8), where B is the associated Brownian motion and N is the Poisson random measure defined by the jumps of L. In this section $(\mathcal{F}_t)_{t\geq 0}$ is the filtration generated by B and N.

Motivated by the representation (9.8), we define the following class of processes.

Definition 9.5.1 We say that a stochastic process $X = (X_t)_{t \geq 0}$ is an *Itô–Lévy process* if it can be written in the form

$$X_t = X_0 + \int_0^t \sigma_s dB_s + \int_0^t b_s ds + \int_0^t \int_{\mathbb{R}_0} c(s, z) \hat{N}(ds, dz), \qquad (9.13)$$

where $\sigma : \Omega \times \mathbb{R}_+ \to \mathbb{R}$, $b : \Omega \times \mathbb{R}_+ \to \mathbb{R}$, and $c : \Omega \times \mathbb{R}_+ \times \mathbb{R}_0 \to \mathbb{R}$ are predictable processes such that, for all $t > 0$,

$$\int_0^t \left(\sigma_s^2 + |b_s| + \int_{\mathbb{R}_0} c^2(s, z) \nu(dz) \right) ds < +\infty \quad \text{a.s.,}$$

and X_0 is an $\mathcal{F}_0$-measurable random variable.

The next result gives Itô's formula for an Itô–Lévy process.

Theorem 9.5.2 (Itô's formula) *Let* $f : \mathbb{R}_+ \times \mathbb{R} \to \mathbb{R}$ *be a function of class* $C^{1,2}$. *Suppose that X is an Itô–Lévy process of the form (9.13). Then, the process $Y_t = f(t, X_t)$ is also an Itô–Lévy process with representation*

$$Y_t = f(0, X_0) + \int_0^t \frac{\partial f}{\partial t}(s, X_s) ds + \int_0^t \frac{\partial f}{\partial x}(s, X_{s-}) dX_s$$

$$+ \frac{1}{2} \int_0^t \frac{\partial^2 f}{\partial x^2}(s, X_s) \sigma_s^2 ds$$

$$+ \sum_{0 \leq s \leq t} \left(f(s, X_s) - f(s, X_{s-}) - \Delta X_s \frac{\partial f}{\partial x}(s, X_{s-}) \right), \qquad (9.14)$$

which holds a.s. for all $t \geq 0$.

Proof For simplicity, we assume that f does not depend on time. The time-dependent case could be proved using similar arguments. The proof is divided into different steps.

Step 1 We first assume that X has the form

$$X_t = X_0 + \int_0^t \int_{\{|z| \geq 1\}} c(s, z) N(ds, dz),$$

where $c : \Omega \times \mathbb{R}_+ \times \mathbb{R}_0 \to \mathbb{R}$ is a predictable process and f is a function in $C(\mathbb{R})$. Let $Z_t = \int_{\{|z| \geq 1\}} z N(t, dz)$. The jump times of the process Z can be defined recursively as $T_0 = 0$ and, for each integer $n \geq 1$,

$$T_n = \inf \{ t > T_{n-1} : |\Delta Z_t| \geq 1 \},$$

where $\Delta Z_t = Z_t - Z_{t-}$. Then, we can write

$$f(X_t) - f(X_0) = \sum_{0 \leq s \leq t} f(X_s) - f(X_{s-}) = \sum_{j=0}^{\infty} f(X_{t \wedge T_{j+1}}) - f(X_{t \wedge T_j})$$

$$= \sum_{j=0}^{\infty} f\left(X_{t \wedge T_{j+1}-} + c(t \wedge T_{j+1}, \Delta Z_{t \wedge T_{j+1}})\right) - f(X_{t \wedge T_{j+1}-}).$$

Hence, we obtain the formula

$$f(X_t) = f(X_0) + \int_0^t \int_{|z| \geq 1} (f(X_{s-} + c(s, z)) - f(X_{s-})) N(ds, dz), \quad (9.15)$$

which coincides with (9.14).

Step 2   We next consider the case where X can be written as

$$X_t = X_0 + \int_0^t \sigma_s dB_s + \int_0^t b_s ds + \int_0^t \int_{\{|z| \geq 1\}} c(s, z) N(ds, dz),$$

where c is as in Step 1 and b and σ are as above. Let $f \in C^2(\mathbb{R})$. Using the stopping times introduced in Step 1, we get

$$f(X_t) - f(X_0) = \sum_{j=0}^{\infty} \left(f(X_{t \wedge T_{j+1}}) - f(X_{t \wedge T_j}) \right)$$

$$= \sum_{j=0}^{\infty} \left(f(X_{t \wedge T_{j+1}-}) - f(X_{t \wedge T_j}) \right) + \sum_{j=0}^{\infty} \left(f(X_{t \wedge T_{j+1}}) - f(X_{t \wedge T_{j+1}-}) \right).$$

Observe that, when $t \in (T_j, T_{j+1})$, the increment that appears in the first sum involves only the continuous part of the process, while the second involves only the jump part. Then, applying Itô's formula for the continuous case (Theorem 2.4.3) to the first sum and formula (9.15) to the second, we obtain

$$f(X_t) = f(X_0) + \int_0^t f'(X_s) \sigma_s dB_s + \int_0^t f'(X_s) b_s ds + \frac{1}{2} \int_0^t f''(X_s) \sigma_s^2 ds$$

$$+ \int_0^t \int_{\{|z| \geq 1\}} (f(X_{s-} + c(s, z)) - f(X_{s-})) N(ds, dz), \quad (9.16)$$

which again gives us (9.14).

Step 3   Suppose now that X is an Itô–Lévy process of the form (9.13). Let $f \in C^2(\mathbb{R})$ with bounded first and second derivatives. Notice that (9.16) holds if we replace $\{|z| \geq 1\}$ by $\{|z| \geq 1/n\}$ for each $n \geq 1$.

Consider the sequence of processes $(X_t^{(n)})_{n \geq 0}$ defined by

$$X_t^{(n)} = X_0 + \int_0^t \sigma_s dB_s + \int_0^t b_s ds + \int_0^t \int_{\{|z| \geq 1/n\}} c(s, z) \hat{N}(ds, dz).$$

Observe that

$$X_t^{(n)} = X_0 + \int_0^t \sigma_s dB_s + \int_0^t \left(b_s - \int_{\{|z| \geq 1/n\}} c(s, z) \nu(dz) \right) ds$$

$$+ \int_0^t \int_{\{|z| \geq 1/n\}} c(s, z) N(ds, dz).$$

Then, by formula (9.16), for each $n \geq 1$,

$$f(X_t^{(n)}) = f(X_0) + \int_0^t f'(X_s^{(n)}) \sigma_s dB_s + \int_0^t f'(X_s^{(n)}) b_s ds$$

$$- \int_0^t \int_{\{|z| \geq 1/n\}} f'(X_{s-}^{(n)}) c(s, z) \nu(dz) ds + \frac{1}{2} \int_0^t f''(X_s^{(n)}) \sigma_s^2 ds$$

$$+ \int_0^t \int_{\{|z| \geq 1/n\}} \left(f(X_{s-}^{(n)} + c(s, z)) - f(X_{s-}^{(n)}) \right) N(ds, dz).$$

We can rewrite the last expression as

$$f(X_t^{(n)}) - f(X_0)$$

$$= \int_0^t f'(X_s^{(n)}) \sigma_s dB_s + \int_0^t f'(X_s^{(n)}) b_s ds + \frac{1}{2} \int_0^t f''(X_s^{(n)}) \sigma_s^2 ds$$

$$+ \int_0^t \int_{\{|z| \geq 1/n\}} \left(f(X_{s-}^{(n)} + c(s, z)) - f(X_{s-}^{(n)}) \right) \hat{N}(ds, dz)$$

$$+ \int_0^t \int_{\{|z| \geq 1/n\}} \left(f(X_{s-}^{(n)} + c(s, z)) - f(X_{s-}^{(n)}) - f'(X_{s-}^{(n)}) c(s, z) \right) \nu(dz) ds.$$

By Corollary 9.4.3, we have that $X_t^{(n)} \to X_t$ in probability, and hence there exists a subsequence that converges to X_t almost surely. Taking the limit along this subsequence, we obtain

$$f(X_t) - f(X_0)$$

$$= \int_0^t f'(X_s) \sigma_s dB_s + \int_0^t f'(X_s) b_s ds + \frac{1}{2} \int_0^t f''(X_s) \sigma_s^2 ds$$

$$+ \int_0^t \int_{\mathbb{R}_0} (f(X_{s-} + c(s, z)) - f(X_{s-})) \hat{N}(ds, dz)$$

$$+ \int_0^t \int_{\mathbb{R}_0} (f(X_{s-} + c(s, z)) - f(X_{s-}) - f'(X_{s-}) c(s, z)) \nu(dz) ds,$$

which implies (9.14). By an approximation argument, this formula also holds for $f \in C^2(\mathbb{R})$. This concludes the proof.   □

9.6 Integral Representation Theorem

In this section we assume that $(\Omega, \mathcal{F}, (\mathcal{F}_t)_{t \geq 0}, \mathrm{P})$ is the filtered probability space of the jump random measure N of a Lévy process. For any $T > 0$, we denote by $L_T^2(\mathcal{P})$ the restriction of $L^2(\mathcal{P})$ to the interval $[0, T]$.

As a consequence of Itô's formula, we have the following representation theorem for square integrable $\mathcal{F}_T$-measurable random variables.

Theorem 9.6.1 (Integral representation theorem)   *Fix $T > 0$, and let F be a random variable in $L^2(\Omega, \mathcal{F}_T, \mathrm{P})$. Then there exists a unique process u in $L_T^2(\mathcal{P})$ such that*

$$F = \mathrm{E}(F) + \int_0^T \int_{\mathbb{R}_0} u(s, z) \hat{N}(ds, dz).$$

Proof   Suppose first that $F = Y_T$, where $Y = (Y_t)_{t \in [0,T]}$ is given by

$$Y_t = \exp\left(\int_0^t \int_{\mathbb{R}_0} h_s z \mathbf{1}_{[0,R]}(z) \hat{N}(ds, dz) \right.$$
$$\left. - \int_0^t \int_{\mathbb{R}_0} \left(\exp\left(h_s z \mathbf{1}_{[0,R]}(z) \right) - 1 - h_s z \mathbf{1}_{[0,R]}(z) \right) \nu(dz) ds \right),$$

where $h \in C([0, T])$ and $R > 0$. The random variable Y_T is called the Doléans–Dade exponential. By Itô's formula (Theorem 9.5.2), it is easy to obtain (see Exercise 9.5)

$$dY_t = Y_{t-} \int_{\mathbb{R}_0} \left(\exp\left(h_t z \mathbf{1}_{[0,R]}(z) \right) - 1 \right) \hat{N}(dt, dz).$$

Hence

$$F = 1 + \int_0^T \int_{\mathbb{R}_0} Y_{s-} \left(\exp\left(h_s z \mathbf{1}_{[0,R]}(z) \right) - 1 \right) \hat{N}(ds, dz),$$

which implies the desired representation because $\mathrm{E}(F) = 1$ and the process $u(s, z) = Y_{s-}(\exp(h_s z \mathbf{1}_{[0,R]}(z)) - 1)$ belongs to $L_T^2(\mathcal{P})$. By linearity, the representation holds for linear combinations of Doléans–Dade exponentials.

In the general case, any random variable $F \in L^2(\Omega, \mathcal{F}_T, \mathrm{P})$ can be approximated in $L^2(\Omega)$ by a sequence F_n of linear combinations of Doléans–Dade exponentials. Then we have

$$F_n = \mathrm{E}(F_n) + \int_0^T \int_{\mathbb{R}_0} u^{(n)}(s, z) \hat{N}(ds, dz),$$

for some sequence $u^{(n)}$ in $L_T^2(\mathcal{P})$. By the isometry of the stochastic integral,

$$E((F_n - F_m)^2) \geq E\left(\left(\int_0^T \int_{\mathbb{R}_0} (u^{(n)}(s,z) - u^{(m)}(s,z))\hat{N}(ds,dz)\right)^2\right)$$

$$= E\left(\int_0^T \int_{\mathbb{R}_0} (u^{(n)}(s,z) - u^{(m)}(s,z))^2 v(dz)ds\right).$$

Hence, $u^{(n)}$ is a Cauchy sequence in $L_T^2(\mathcal{P})$ and it converges to a process u in $L_T^2(\mathcal{P})$. Again applying the isometry property, and taking into account that $E(F_n)$ converges to $E(F)$, we obtain

$$F = \lim_{n \to \infty} F_n = \lim_{n \to \infty} \left(E(F_n) + \int_0^T \int_{\mathbb{R}_0} u^{(n)}(s,z)\hat{N}(ds,dz)\right)$$

$$= E(F) + \int_0^T \int_{\mathbb{R}_0} u(s,z)\hat{N}(ds,dz).$$

Finally, uniqueness follows also from the isometry property. $\square$

Example 9.6.2 Consider the process $X_t = \int_0^t \int_{\mathbb{R}_0} z\,\hat{N}(ds,dz)$, $t \in [0,T]$. We want to find the integral representation of $F = X_T^2$. By Itô's formula (Theorem 9.5.2),

$$X_t^2 = \int_0^t \int_{\mathbb{R}_0} \left((X_{s-} + z)^2 - X_{s-}^2 - 2X_{s-}z\right)v(dz)ds$$

$$+ \int_0^t \int_{\mathbb{R}_0} \left((X_{s-} + z)^2 - X_{s-}^2\right)\hat{N}(ds,dz)$$

$$= t\int_{\mathbb{R}_0} z^2 v(dz) + \int_0^t \int_{\mathbb{R}_0} z(2X_{s-} + z)\hat{N}(ds,dz).$$

Therefore, we obtain the representation

$$F = T\int_{\mathbb{R}_0} z^2 v(dz) + \int_0^T \int_{\mathbb{R}_0} z(2X_{s-} + z)\hat{N}(ds,dz). \qquad (9.17)$$

Corollary 9.6.3 (Martingale representation theorem) *Let $(M_t)_{t \geq 0}$ be a square integrable martingale with respect to $(\mathcal{F}_t)_{t \geq 0}$. Then there exists a unique process u in $L^2(\mathcal{P})$ such that*

$$M_t = E(M_0) + \int_0^t \int_{\mathbb{R}_0} u(s,z)\hat{N}(ds,dz),$$

for all $t \geq 0$.

9.7 Girsanov's Theorem

Assume that $(\Omega, \mathcal{F}, (\mathcal{F}_t)_{t\geq0}, P)$ is the filtered probability space of the jump random measure N of a Lévy process. We next present Girsanov's theorem for the Poisson random measure N (see Lépingle and Mémin, 1978, and Di Nunno *et al.*, 2009, Theorem 12.21).

Let $c: \Omega \times [0, T] \times \mathbb{R}_0 \to \mathbb{R}$ be a predictable process satisfying $c(s, z) > -1$, for all $s \in [0, T]$ and $z \in \mathbb{R}_0$, and

$$\int_0^T \int_{\mathbb{R}_0} (|\log(1 + c(s, z))|^2 + c^2(s, z))v(dz)ds < \infty \quad \text{a.s.} \quad (9.18)$$

For all $t \in [0, T]$, set

$$Z_t = \exp\left(\int_0^t \int_{\mathbb{R}_0} \log(1 + c(s, z))\hat{N}(ds, dz) \right.$$
$$\left. + \int_0^t \int_{\mathbb{R}_0} (\log(1 + c(s, z)) - c(s, z))v(dz)ds \right).$$

By Itô's formula (Theorem 9.5.2), it is easy to obtain that (see Exercise 9.7) Z_t is the unique solution to the stochastic differential equation

$$dZ_t = Z_{t-} \int_{\mathbb{R}_0} c(t, z)\hat{N}(dt, dz), \quad Z_0 = 1.$$

Theorem 9.7.1 *Let* $c: \Omega \times [0, T] \times \mathbb{R}_0 \to \mathbb{R}$ *be a predictable process satisfying* $c(s, z) > -1$, *for all* $s \in [0, T]$ *and* $z \in \mathbb{R}_0$, *and* (9.18). *Assume that*

$$E\left(\exp\left(\int_0^T \int_{\mathbb{R}_0} ((1 + c(s, z))\log(1 - c(s, z)) - c(s, z))v(dz)ds \right) \right) < \infty.$$

Then, for all $t \in [0, T]$, $E(Z_t) = 1$ *and* $(Z_t)_{t\in[0,T]}$ *is a positive martingale.*

By Theorem 9.7.1, the following defines a probability measure Q on $(\Omega, \mathcal{F}_T)$:

$$\frac{dQ}{dP}(\omega) = Z_T(\omega), \quad \omega \in \Omega.$$

We next define the stochastic process

$$\hat{N}_Q(t, A) = \hat{N}(t, A) - v(t, A), \quad A \in \mathcal{A}_v,$$

where

$$v(t, A) = \int_0^t \int_A c(s, z)v(dz)ds.$$

Theorem 9.7.2 (Girsanov's theorem)   *Under the probability measure* Q, *the stochastic process* $\{\hat{N}_Q(t, A), A \in \mathcal{A}_y, t \in [0, T]\}$ *is a compensated Poisson random measure.*

9.8 Multiple Stochastic Integrals

Consider a Poisson random measure M on a complete separable metric space $(Z, \mathcal{Z}, m)$, where m is a σ-finite atomless measure.

Fix $n \geq 2$. We denote by $L^2(Z^n)$ the space of real-valued measurable functions on Z^n that are square integrable with respect to m^n. We write $L_s^2(Z^n)$ to indicate the subspace of $L^2(Z^n)$ composed of symmetric functions. We define the symmetrization $\tilde{h}$ of h in $L^2(Z^n)$ by

$$\tilde{h}(z_1, \ldots, z_n) = \frac{1}{n!} \sum_\sigma h(z_{\sigma(1)}, \ldots, z_{\sigma(n)}),$$

where the sum runs over all permutations σ of $\{1, \ldots, n\}$. By Jensen's inequality, we have

$$\|\tilde{h}\|_{L^2(Z^n)} \leq \|h\|_{L^2(Z^n)}. \tag{9.19}$$

We next define the multiple stochastic integral of a function h in $L^2(Z^n)$ with respect to the compensated Poisson random measure $\hat{M}$.

Let $\mathcal{I}_n$ be the vector space generated by the set of elementary functions on Z^n of the form

$$h(z_1, \ldots, z_n) = \sum_{j_1, \ldots, j_n = 1}^{k} a_{j_1, \ldots, j_n} \mathbf{1}_{A_1 \times \cdots \times A_n}(z_1, \ldots, z_n), \tag{9.20}$$

where each $a_{j_1, \ldots, j_n} \in \mathbb{R}$ is zero whenever at least two indices $j_1, \ldots, j_n$ coincide and $A_1, \ldots, A_n \in \mathcal{Z}$ are pairwise disjoints.

Definition 9.8.1   If h is a function in $\mathcal{I}_n$ of the form (9.20) we define the *multiple stochastic integral* of h with respect to the compensated Poisson random measure $\hat{M}$ by

$$I_n(h) = \sum_{j_1, \ldots, j_n = 1}^{k} a_{j_1, \ldots, j_n} \hat{M}(A_1) \cdots \hat{M}(A_n),$$

It is easy to check that the mapping $h \to I_n(h)$ is linear. Moreover, for each $h \in \mathcal{I}_n$, we have $I_n(h) = I_n(\tilde{h})$ (see Exercise 9.8). Therefore, it suffices to restrict ourselves to the case of symmetric functions. Given $h \in \mathcal{I}_n$ of the form (9.20), it is easy to check that h is symmetric if and only if, for each permutation σ of $\{1, \ldots, n\}$ and $1 \le j_1, \ldots, j_n \le k$, $a_{j_1, \ldots, j_n} = a_{j_{\sigma(1)}, \ldots, j_{\sigma(n)}}$. We denote by I_n^s the set of functions in I_n that are symmetric.

The multiple stochastic integral has the following isometry property (see Appelbaum, 2009, Theorem 5.4.4).

Theorem 9.8.2 *Let $h \in I_n^s$ and $g \in I_k^s$, $n, k \ge 1$. Then* $E(I_n(h)) = 0$ *and*

$$E(I_n(h)I_k(g)) = n! \langle h, g \rangle_{L^2(Z^n)} \mathbf{1}_{\{n=k\}}. \tag{9.21}$$

Now, since the vector space I_n is dense in $L^2(Z^n)$ (see Appelbaum, 2009, Appendix 5.9), if $h \in L^2(Z^n)$ then there exists a sequence $(h_\ell)_{\ell \ge 1} \in I_n$ that converges to h in $L^2(Z^n)$ as $\ell \to \infty$. Using the above isometry property (Theorem 9.8.2) and inequality (9.19), we obtain

$$E((I_n(h_p) - I_n(h_q))^2) = n! \|\tilde{h}_p - \tilde{h}_q\|_{L^2(Z^n)}^2 \le n! \|h_p - h_q\|_{L^2(Z^n)}^2 \to 0,$$

as $p, q \to \infty$. Therefore, the sequence $(I_n(h_\ell))_{\ell \ge 1}$ is Cauchy in $L^2(\Omega)$ and is thus convergent. The multiple stochastic integral of h in $L^2(Z^n)$ is the limit of the sequence $\{I_n(h_\ell)\}_{\ell \ge 1}$ in $L^2(\Omega)$ and is denoted

$$I_n(h) = \int_{Z^n} h(z_1, \ldots, z_n) \, \hat{M}(dz_1) \cdots \hat{M}(dz_n).$$

We observe that $I_1(h) = \hat{M}(h)$. Moreover, Theorem 9.8.2 also holds for all $h \in L^2(Z^n)$ and $g \in L^2(Z^k)$.

The next result shows that the multiple stochastic integral of the jump measure of a Lévy process coincides with the iterated integral.

Proposition 9.8.3 (Appelbaum, 2009, Theorem 5.4.5) *Let N be the jump measure of a Lévy process. Then, for any $h \in L^2((\mathbb{R}_+ \times \mathbb{R}_0)^n)$,*

$$I_n(h) = n! \int_{\mathbb{R}_+} \int_{\mathbb{R}_0} \cdots \int_0^{t_2^-} \int_{\mathbb{R}_0} h(t_1, z_1, \ldots, t_n, z_n) \hat{N}(dt_1, dz_1) \cdots \hat{N}(dt_n, dz_n).$$

We next give a formula for the product of two multiple stochastic integrals that was proved by, for example, Kabanov (1975) and Surgailis (1984). Let $f \in L_s^2(Z^n)$ and $g \in L_s^2(Z^k)$. For any $r = 0, \ldots, n \wedge k$ and $\ell = 1, \ldots, r$, we define the *modified contraction* $f \star_r^\ell g$ of f and g to be the element of $Z^{n+k-r-\ell}$ given by

$$\left(f \star_r^\ell g \right)(\gamma_1, \ldots, \gamma_{r-\ell}, t_1, \ldots, t_{n-r}, s_1, \ldots, s_{k-r})$$

$$= \int_{Z^\ell} f(z_1, \ldots, z_\ell, \gamma_1, \ldots, \gamma_{r-\ell}, t_1, \ldots, t_{n-r})$$

$$\times g(z_1, \ldots, z_\ell, \gamma_1, \ldots, \gamma_{r-\ell}, s_1, \ldots, s_{k-r}) m^\ell(dz_1 \cdots dz_\ell)$$

and, for $\ell = 0$,

$$\left(f \star_r^0 g\right)(\gamma_1, \ldots, \gamma_r, t_1, \ldots, t_{n-r}, s_1, \ldots, s_{k-r})$$
$$= f(\gamma_1, \ldots, \gamma_r, t_1, \ldots, t_{n-r}) g(\gamma_1, \ldots, \gamma_r, s_1, \ldots, s_{k-r}),$$

so that $\left(f \star_0^0 g\right)(t_1, \ldots, t_n, s_1, \ldots, s_k) = f(t_1, \ldots, t_n) g(s_1, \ldots, s_k)$.

When $\ell = r$, $f \star_r^r g$ belongs to $L^2(Z^{n+k-2r})$. Otherwise, there are cases where the integrals are not even well defined. We denote by $f\widetilde{\star}_r^\ell g$ the symmetrization of $f \star_r^\ell g$. Then, the product of two multiple stochastic integrals satisfies the following formula: suppose that, for all $r = 0, \ldots, n \wedge k$ and $\ell = 1, \ldots, r$, $f \star_r^\ell g$ belongs to $L^2(Z^{n+k-r-\ell})$. Then

$$I_n(f)I_k(g) = \sum_{r=0}^{n \wedge k} r! \binom{n}{r}\binom{k}{r} \sum_{\ell=0}^{r} \binom{r}{\ell} I_{n+k-r-\ell}\left(f\widetilde{\star}_r^\ell g\right). \tag{9.22}$$

9.9 Wiener Chaos for Poisson Random Measures

Consider a Poisson random measure M on a complete separable metric space $(Z, \mathcal{Z}, m)$, where m is a σ-finite atomless measure. Recall that $\hat{M}$ denotes the compensated Poisson random measure. We assume in this section that $\mathcal{F}$ is the σ-field generated by N and the P-null sets.

Definition 9.9.1 The *nth Wiener chaos* associated with $\hat{M}$ is denoted $\mathcal{H}_n$ and defined as the Hilbert space of random variables of the type $I_n(h)$, where $n \geq 1$ and $h \in L_s^2(Z^n)$; that is, $\mathcal{H}_n = I_n(L_s^2(Z^n))$. Moreover, we define $\mathcal{H}_0$ as $\mathbb{R}$.

Note that $\mathcal{H}_1 = \{\hat{M}(h), h \in L^2(Z)\}$. Observe that, by the isometry property, for all $n \geq 1$, $\mathcal{H}_n$ is a closed linear subspace of $L^2(\Omega)$ and the $\mathcal{H}_n$ are all orthogonal. The following chaotic representation property is due to Itô (1956, Theorem 2).

Theorem 9.9.2 *We have that $L^2(\Omega) = \oplus_{n \geq 0} \mathcal{H}_n$; that is, any random variable $F \in L^2(\Omega)$ admits the following unique decomposition*

$$F = \sum_{n=0}^{\infty} I_n(h_n), \tag{9.23}$$

where $h_0 = \mathrm{E}(F)$, I_0 is the identity mapping on constants and $h_n \in L_s^2(Z^n)$.

Proof The proof is divided into different steps.

Step 1   First observe that the set of simple random variables

$$S = \left\{ \varphi(M(A_1), \ldots, M(A_n)), n \geq 1, \varphi \in C_b(\mathbb{R}^n), A_i \in \mathcal{Z}_m \text{ pairwise disjoint} \right\}$$

is dense in $L^2(\Omega)$.

Step 2   Next we show that the set

$$\mathcal{P} = \left\{ M(A_1)^{p_1} \cdots M(A_n)^{p_n}, n \geq 1, p_i \geq 0, A_i \in \mathcal{Z}_m \text{ pairwise disjoint} \right\}$$

is dense in $L^2(\Omega)$. First observe that any element in this set belongs to
$L^2(\Omega)$ since it has a finite norm. Now, by Step 1, it suffices to show that if
F belongs to S and $E(FG) = 0$ for all $G \in \mathcal{P}$ then $F = 0$ a.s. Let F be in S
and, to simplify the exposition, we assume that $n = 1$; that is, $F = \varphi(M(A))$,
$\varphi \in C_b(\mathbb{R})$. Let σ denote the Poisson distribution with parameter $m(A)$, and
assume that, for all $p \geq 0$,

$$E(FM(A)^p) = \int_{\mathbb{R}} \varphi(x) x^p \sigma(dx) = 0.$$

Furthermore, for all $t \in \mathbb{R}$,

$$\int_{\mathbb{R}} e^{|tx|} |\varphi(x)| \, \sigma(dx) \leq \left(\int_{\mathbb{R}} e^{2|tx|} \sigma(dx) \right)^{1/2} \left(\int_{\mathbb{R}} |\varphi(x)|^2 \, \sigma(dx) \right)^{1/2},$$

which is finite. Thus, our assumption implies that, for all $t \in \mathbb{R}$,

$$\int_{\mathbb{R}} e^{itx} \varphi(x) \, \sigma(dx) = 0,$$

from which we conclude that $\varphi(M(A)) = 0$ a.s.

Step 3   Next consider the set

$$Q = \left\{ M(A_1) \cdots M(A_n), n \geq 1, A_i \in \mathcal{Z}_m \text{ pairwise disjoint} \right\}.$$

We claim that Q is dense in $L^2(\Omega)$. By Step 2, it suffices to show that any
element in $\mathcal{P}$ belongs to the closed linear span of the family Q. Let $M \in \mathcal{P}$
be of the form $M(A_1)^{p_1} \cdots M(A_n)^{p_n}$, $n \geq 1$, $p_i \geq 0$, with the $A_i \in \mathcal{Z}_m$
pairwise disjoint. Consider a subdivision $\{B_1, \ldots, B_k\}$ of $\{A_1, \ldots, A_n\}$ such
that $m(B_i) < \min(\epsilon / m(B_1 \cup \cdots \cup B_k), 1)$ for some $\epsilon > 0$. Then,

$$M = \sum M(B_{i_1})^{j_1} \cdots M(B_{i_r})^{j_r},$$

with $i_1 < \cdots < i_r$. Since $M(B_i)$ takes only nonnegative integer values, we
have

$$M \geq \sum M(B_{i_1}) \cdots M(B_{i_r}) = M_\epsilon$$

and

$$P(M \neq M_\epsilon) = P(M(B_i) > 2 \text{ for some } i)$$

$$\leq \sum_{i=1}^{k} P(M(B_i) \geq 2) \leq \sum_{i=1}^{k} m(B_i)^2 < \epsilon.$$

Therefore M_ϵ converges to M in probability as ϵ goes to 0. Furthermore, we have $0 \leq M_\epsilon \leq M$ and $M \in L^2(\Omega)$. Therefore, M_ϵ converges to M in $L^2(\Omega)$ as ϵ goes to 0, and the claim is proved.

Step 4   Finally, it suffices to check that, for any $n \geq 1$ and $A_1, \ldots, A_n \in \mathcal{Z}_m$ pairwise disjoint,

$$M(A_1) \cdots M(A_n) = m(A_1) \cdots m(A_m) + I_1(h_1) + \cdots + I_n(h_n),$$

for some $h_i \in L^2_s(Z^i)$. Consider the function

$$h(z_1, \ldots, z_n) = \mathbf{1}_{A_1 \times \cdots \times A_n}(z_1, \ldots, z_n).$$

Then

$$I_n(h) = \hat{M}(A_1) \cdots \hat{M}(A_n) = (M(A_1) - m(A_1)) \cdots (M(A_n) - m(A_n))$$
$$= M(A_1) \cdots M(A_n) + R + m(A_1) \cdots m(A_m),$$

where R is a linear combination of elements of the form $M(A_{i_1}) \cdots M(A_{i_p})$, with $1 \leq p < n$. Thus the result follows by induction on $n \geq 1$.   □

Example 9.9.3   Consider the process $X_t = \int_0^t \int_{\mathbb{R}_0} z \hat{N}(ds, dz)$, $t \in [0, T]$. By (9.17) and Proposition 9.8.3,

$$X_T^2 = T \int_{\mathbb{R}_0} z^2 \nu(dz) + \int_0^T \int_{\mathbb{R}_0} z(2X_{s-} + z)\hat{N}(ds, dz)$$

$$= T \int_{\mathbb{R}_0} z^2 \nu(dz) + \int_0^T \int_{\mathbb{R}_0} z^2 \hat{N}(ds, dz)$$

$$+ \int_0^T \int_{\mathbb{R}_0} \int_0^{t_2-} \int_{\mathbb{R}_0} 2z_1 z_2 \hat{N}(ds_1, dz_1)\hat{N}(ds_2, dz_2)$$

$$= \sum_{n=0}^{2} I_n(h_n),$$

where $h_0 = T \int_{\mathbb{R}_0} z^2 \nu(dz)$, $h_1(s_1, z_1) = z_1$ and $h_2(s_1, z_1, s_2, z_2) = z_1 z_2$.

Exercises

9.1 Let $(\tau_i)_{i \geq 1}$ be a sequence of independent exponential random variables with parameter $\lambda > 0$. Set $T_n = \sum_{i=1}^{n} \tau_i$ and $N_t = \sum_{n \geq 1} 1_{\{t \geq T_n\}}$. Show that for any $t > 0$, N_t has a Poisson distribution with parameter λt.

9.2 Let N be a Poisson random measure associated with a Lévy process. Show that the process $(\hat{N}(t, A))_{t \geq 0}$ is an $\mathcal{F}_t$-martingale, for any $A \in \mathcal{A}_\nu$.

9.3 Let $u \in L^2_{loc}(\mathcal{P})$. Show that there exists a sequence of processes $u^{(n)}$ in $\mathcal{E}$ such that, for all $T > 0$, the sequence

$$\int_0^T \int_{\mathbb{R}_0} |u(t, x) - u^{(n)}(t, z)|^2 \nu(dz) dt$$

converges to zero in probability.

9.4 Show that Proposition 9.4.2 is also true for all $u \in L^2_{loc}(\mathcal{P})$.

9.5 Consider the process $Y = (Y_t)_{t \in [0,T]}$ given by

$$Y_t = \exp\left(\int_0^t \int_{\mathbb{R}_0} h_s z \hat{N}(ds, dz) \right.$$
$$\left. - \int_0^t \int_{\mathbb{R}_0} (e^{h_s z} - 1 - h_s z) \nu(dz) ds \right),$$

where $h \in L^2([0, T])$ and is càdlàg. Show that

$$dY_t = Y_{t-} \int_{\mathbb{R}_0} (e^{h_t z} - 1) \hat{N}(dt, dz).$$

In particular, Y is a local martingale.

9.6 Consider the process $X_t = \int_0^t \int_{\mathbb{R}_0} z \hat{N}(ds, dz)$, $t \in [0, T]$. Find the integral representation of the following random variables:

$$X_T^3, \quad e^{X_T}, \quad \sin X_T, \quad \int_0^T X_t dt.$$

9.7 Let $c: \Omega \times [0, T] \times \mathbb{R}_0 \to \mathbb{R}$ be a predictable process satisfying $c(s, z) < 1$, for all $s \in [0, T]$ and $z \in \mathbb{R}_0$, and also satisfying

$$\int_0^T \int_{\mathbb{R}_0} (|\log(1 - c(s, z))|^2 + c^2(s, z)) \nu(dz) ds < +\infty \quad \text{a.s.}$$

For all $t \in [0, T]$, set

$$Z_t = \exp\left(\int_0^t \int_{\mathbb{R}_0} \log(1 - c(s, z)) \hat{N}(ds, dz) \right.$$
$$\left. + \int_0^t \int_{\mathbb{R}_0} (\log(1 - c(s, z)) + c(s, z)) \nu(dz) ds \right).$$

Show that

$$dZ_t = Z_{t-} \int_{\mathbb{R}_0} c(t, z) \hat{N}(dt, dz), \quad Z_0 = 1.$$

9.8   Show that, for all $h \in I_n$, we have $I_n(h) = I_n(\tilde{h})$, where $\tilde{h}$ denotes the symmetrization of h.

9.9   Find the chaos expansion of the random variable Y_T, where Y is as in Exercise 9.5.

9.10 Find the chaos expansion of the random variables in Exercise 9.6.

10

Malliavin Calculus for Jump Processes I

In this chapter we develop the Malliavin calculus in the Poisson framework using the Wiener chaos decomposition. We start by defining the Malliavin derivative, which in this case is a closed operator satisfying a chain rule different from that obtained for Brownian motion. We then define the divergence operator and show that, when acting on square integrable and predictable processes, it coincides with the stochastic integral with respect to the Poisson random measure associated with a Lévy process. As in the Wiener case, we define the Ornstein–Uhlenbeck semigroup and its generator. Then we present the jump version of the Clark–Ocone formula. Finally, we combine Stein's method with Malliavin calculus to study normal approximations in the Poisson framework.

10.1 Derivative Operator

Consider a Poisson random measure M on a complete separable metric space $(Z, \mathcal{Z}, m)$, where m is a σ-finite atomless measure. Let $L_s^2(Z^n)$ be the space of symmetric square integrable functions on Z^n. Given $h \in L_s^2(Z^n)$ and fixed $z \in Z$, we write $h(\cdot, z)$ to indicate the function on Z^{n-1} given by $(z_1, \ldots, z_{n-1}) \rightarrow h(z_1, \ldots, z_{n-1}, z)$.

Definition 10.1.1 We denote by $\mathbb{D}^{1,2}$ the set of random variables F in $L^2(\Omega)$ with a chaotic decomposition $F = \sum_{n=0}^{\infty} I_n(h_n)$, $h_n \in L_s^2(Z^n)$, satisfying

$$\sum_{n \geq 1} nn! \|h_n\|_{L^2(Z^n)}^2 < \infty. \tag{10.1}$$

Then, if $F \in \mathbb{D}^{1,2}$, we define the *Malliavin derivative D* of F as the $L^2(Z)$-valued random variable given by

$$D_z F = \sum_{n \geq 1} n I_{n-1}(h_n(\cdot, z)), \quad z \in Z.$$

For example, if $F = I_1(h)$ then $DF = h$ and if $F = I_2(h)$ then $D_z F = 2I_1(f(\cdot, z))$. Observe that if F belongs to $\mathbb{D}^{1,2}$ then

$$\mathrm{E}\left(\|DF\|^2_{L^2(Z)}\right) = \sum_{n \geq 1} nn! \|h_n\|^2_{L^2(Z^n)}.$$

It is easy to see that $\mathbb{D}^{1,2}$ is a Hilbert space with respect to the scalar product

$$\langle F, G \rangle_{1,2} = \mathrm{E}(FG) + \mathrm{E}(\langle DF, DG \rangle_{L^2(Z)}),$$

where

$$\langle DF, DG \rangle_{L^2(Z)} = \int_Z D_z F D_z G\, m(dz);$$

this defines the following seminorm in $\mathbb{D}^{1,2}$

$$\|F\|_{1,2} = \left(\mathrm{E}(|F|^2) + \mathrm{E}\left(\|DF\|^2_{L^2(Z)}\right)\right)^{1/2}.$$

We identify the space $L^2(\Omega; L^2(Z))$ with $L^2(\Omega \times Z)$. The next result shows that D is a closed operator from $\mathbb{D}^{1,2} \subset L^2(\Omega)$ into $L^2(\Omega \times Z)$.

Proposition 10.1.2 *Let $(F_k)_{k \geq 1}$ be a sequence of random variables in $\mathbb{D}^{1,2}$ that converges to F in $L^2(\Omega)$ and is such that the sequence $(DF_k)_{k \geq 1}$ converges to u in $L^2(\Omega \times Z)$. Then F belongs to $\mathbb{D}^{1,2}$ and $DF = u$.*

Proof Set $F = \sum_{n=0}^\infty I_n(h_n)$ and $F_k = \sum_{n=0}^\infty I_n(h_n^k)$, with $h_n, h_n^k \in L^2_s(Z^n)$. Then h_n^k converges to h_n in $L^2(Z^n)$. Furthermore, by Fatou's lemma,

$$\lim_{k \to \infty} \sum_{n=1}^\infty nn! \|h_n^k - h_n\|^2_{L^2(Z^n)} \leq \lim_{k \to \infty} \liminf_{j \to \infty} \sum_{n=1}^\infty nn! \|h_n^k - h_n^j\|^2_{L^2(Z^n)} = 0,$$

which implies the desired statement. $\qquad\qquad\qquad\qquad\qquad\square$

Given $z \in Z$ fixed, and a point measure ω in Ω, we define the map

$$\varphi_z(\omega) = \omega + \delta_z.$$

Observe that the map φ_z is well defined from Ω to Ω. By (9.5), if we set $\Omega_0 = \{\omega \in \Omega : \omega(\{\alpha\}) \leq 1, \forall \alpha \in Z\}$ then $P(\Omega_0) = 1$. Moreover, because $P(\omega : \omega(\{z\}) = 1) = 0$, we have that $\varphi_z(\Omega_0) \subset \Omega_0$ a.s. for every fixed $z \in Z$. The map φ_z induces a transformation on random variables $F \colon \Omega \to \mathbb{R}$ defined by

$$(\varphi_z(F))(\omega) = F(\omega + \delta_z) - F(\omega).$$

The next result gives a product rule for the map φ_z, and it will be used later.

Lemma 10.1.3 *Let $F, G: \Omega \to \mathbb{R}$ be random variables. Then*

$$\varphi_z(FG) = F\varphi_z(G) + G\varphi_z(F) + \varphi_z(F)\varphi_z(G).$$

Proof We have

$$\varphi_z(FG) = F(\omega + \delta_z)G(\omega + \delta_z) - F(\omega)G(\omega)$$

$$= F(\omega + \delta_z)G(\omega + \delta_z) - F(\omega + \delta_z)G(\omega) + F(\omega + \delta_z)G(\omega) - F(\omega)G(\omega)$$

$$= F(\omega + \delta_z)\varphi_z(G) + \varphi_z(F)G(\omega)$$

$$= (F(\omega + \delta_z) - F(\omega))\,\varphi_z(G) + \varphi_z(F)G + F\varphi_z(G)$$

$$= F\varphi_z(G) + G\varphi_z(F) + \varphi_z(F)\varphi_z(G),$$

which concludes the proof. □

The next result shows that φ_z and D_z coincide.

Theorem 10.1.4 *Let F be a random variable in $L^2(\Omega)$. Then F belongs to $\mathbb{D}^{1,2}$ if and only if there exists a version of F such that the stochastic process $z \to \varphi_z(F)$ belongs to $L^2(\Omega \times Z)$. In this case*

$$D_z F = \varphi_z(F), \qquad (10.2)$$

for almost all $(\omega, z) \in \Omega \times Z$.

Proof We divide the proof into different steps.

Step 1 Assume first that $F = I_1(h)$. Then $D_z F = h(z)$. Thus, (10.2) is just a consequence of the following relation:

$$F_z - F = \int_Z h(\alpha)(\hat{M}(d\alpha) + \delta_z) - \int_Z h(\alpha)\hat{M}(d\alpha) = \int_Z h(\alpha)\delta_z = h(z).$$

Step 2 Consider now an elementary function h of the form (9.20). Then

$$D_z(I_n(h)) = \sum_{j_1,\ldots,j_n=1}^{m} a_{j_1,\ldots,j_n} \sum_{i=1}^{n} \hat{M}(A_1) \cdots \mathbf{1}_{A_i}(z) \cdots \hat{M}(A_n).$$

Moreover,

$$\varphi_z(I_n(h)) = \sum_{j_1,\ldots,j_n=1}^{m} a_{j_1,\ldots,j_n} \Big((\hat{M} + \delta_z)(A_1) \cdots (\hat{M} + \delta_z)(A_n)$$

$$- \hat{M}(A_1) \cdots \hat{M}(A_n) \Big).$$

Now, observe that if $z \notin A_i$ for all $i = 1, \ldots, n$ then $\varphi_z(I_n(h)) = 0$, while if $z \in A_i$ for some $i = 1, \ldots, n$ then

$$\varphi_z(I_n(h)) = \sum_{j_1,\ldots,j_n=1}^{m} a_{j_1,\ldots,j_n} \hat{M}(A_1) \cdots \hat{M}(A_{i-1}) \hat{M}(A_{i+1}) \cdots \hat{M}(A_n).$$

Therefore, (10.2) holds if $F = I_n(h)$, where h is of the form (9.20), that is, $h \in \mathcal{I}_n$.

Step 3   Consider a random variable F of the form $I_n(h)$ where h is in $L^2(Z^n)$. Then F is the limit in $L^2(\Omega)$ of a sequence $F^k = I_n(h^k)$, where the h_k belong to $\mathcal{I}_n$. Therefore F is the almost sure limit of a subsequence $(F^{k_j})_{j \geq 0}$. Consequently, $\varphi_z(F)$ is the almost sure limit of $\varphi_z(F^{k_j})$ for almost every z. Moreover, DF^{k_j} converges to DF in $L^2(\Omega \times Z)$ as $j \to \infty$, and (9.20) holds.

Step 4   We now assume that $F \in L^2(\Omega)$ with chaotic decomposition (9.23). Then F is the limit in $L^2(\Omega)$ of the sequence of partial sums $F_k = \sum_{n=0}^{k} I_n(h_n)$. Thus, again F is the a.s. limit of a subsequence $(F_{k_j})_{j \geq 0}$ and $\varphi_z(F)$ is the a.s. limit of $\varphi_z(F_{k_j})$ as $j \to \infty$ for almost all z. Since D is closed, the desired statement follows. □

In Nualart and Vives (1990), the authors studied the derivative and divergence operators in terms of the Wiener chaos in an arbitrary Fock space associated with a Hilbert space. In particular, they considered the Wiener and Poisson spaces. The next example is taken from there.

Example 10.1.5   As an application of Theorem 10.1.4, we will compute the Malliavin derivative of the jump times of a Poisson process. Let $Z = [0, 1]$ and let $T_1, \ldots, T_n$ be the jump times of a Poisson process $(N_t)_{t \in [0,1]}$. We need to compute the transformation $\varphi_t(T_i) = T_i(\omega + \delta_t) - T_i(\omega)$ for all $i = 1, \ldots, n$. We have $\varphi_t(T_i) = 0$ if $t > T_i$, $\varphi(T_i) = t - T_i$ if $T_{i-1} < t < T_i$, and $\varphi_t(T_i) = T_{i-1} - T_i$ if $t < T_{i-1}$. Therefore

$$\varphi_t(T_i) = T_{i-1} \mathbf{1}_{\{t < T_i\}} + t \mathbf{1}_{\{T_{i-1} < t < T_i\}} - T_i \mathbf{1}_{\{t < T_i\}}.$$

For example for $i = 1$, $\varphi_t(T_1) = (t - T_1) \mathbf{1}_{[0,T_1]}(t)$.

Remark 10.1.6   Observe that the Malliavin derivative D is not a local operator as in the Brownian motion case (see Lemma 3.4.1). For example, with the notation of Example 10.1.5, let F be a random variable in $\mathbb{D}^{1,2}$ that is equal to zero over the set $\{N_{1/2} = N_1\}$ and takes the value 1 on the complementary set. Then, for all $t > 1/2$, $D_t F = 1$.

As a consequence of Lemma 10.1.3, we have the following product rule for the Malliavin derivative.

Lemma 10.1.7 *Let $F, G \in \mathbb{D}^{1,2}$. Suppose that $FG \in L^2(\Omega)$ and $(F + DF)(G + DG) \in L^2(\Omega \times Z)$. Then the product FG also belongs to $\mathbb{D}^{1,2}$ and*

$$D(FG) = FDG + GDF + DFDG.$$

We can also prove the following chain rule (see Di Nunno *et al.*, 2009).

Proposition 10.1.8 *Let F be a random variable in $\mathbb{D}^{1,2}$ and let φ be a real continuous function such that $\varphi(F)$ belongs to $L^2(\Omega)$ and $\varphi(F + DF)$ belongs to $L^2(\Omega \times Z)$. Then $\varphi(F)$ belongs to $\mathbb{D}^{1,2}$ and*

$$D\varphi(F) = \varphi(F + DF) - \varphi(F).$$

Proof First assume that φ has compact support and F is bounded. Recall that, from the Fourier inversion theorem,

$$\varphi(F) = \frac{1}{\sqrt{2\pi}} \int_{\mathbb{R}} e^{iyF} \hat{\varphi}(y) dy,$$

where $\hat{\varphi}$ denotes the Fourier transform of φ, that is,

$$\hat{\varphi}(y) = \frac{1}{\sqrt{2\pi}} \int_{\mathbb{R}} e^{-ixy} \varphi(x) dx.$$

Then, using Proposition 10.1.2 and Exercise 10.1, we get

$$D\varphi(F) = \frac{1}{\sqrt{2\pi}} \int_{\mathbb{R}} \sum_{n=0}^{\infty} \frac{1}{n!} (iy)^n ((F + DF)^n - F^n) \hat{\varphi}(y) dy$$

$$= \frac{1}{\sqrt{2\pi}} \int_{\mathbb{R}} \left(e^{iy(F+DF)} - e^{iyF} \right) \hat{\varphi}(y) dy$$

$$= \varphi(F + DF) - \varphi(F).$$

Hence, the result holds in this case. For a general continuous function φ and random variables F satisfying the assumptions of the proposition, the result follows by approximation and using the fact that D is closed. □

Example 10.1.9 Consider the process $X_t = \int_0^t \int_{\mathbb{R}_0} z \hat{N}(ds, dz)$, $t \in [0, T]$. By Example 9.9.3,

$$D_{t,z} X_T^2 = z^2 + 2I_1(h_2(\cdot, t, z)) = z^2 + 2X_T z. \tag{10.3}$$

Furthermore, since $D_{t,z} X_T = z$, using Proposition 10.1.8 we obtain

$$D_{t,z} X_T^2 = (X_T + z)^2 - X_T^2 = z^2 + 2X_T z,$$

which coincides with (10.3).

10.2 Divergence Operator

Observe that, owing to the chaotic representation (9.23), every stochastic process u in $L^2(\Omega \times Z)$ admits a unique representation of the form

$$u(z) = \sum_{n \geq 0} I_n(h_n(\cdot, z)), \tag{10.4}$$

where, for each $z \in Z$, the function $h_n(\cdot, z)$ belongs to $L^2_s(Z^n)$ and

$$\|u\|^2_{L^2(\Omega \times Z)} = \sum_{n \geq 0} n! \|h_n\|^2_{L^2(Z^{n+1})} < \infty.$$

Definition 10.2.1 The *domain* of the divergence operator δ, denoted by Dom δ, is defined as the set of stochastic processes u in $L^2(\Omega \times Z)$ such that the chaotic expansion (10.4) verifies the condition

$$\sum_{n \geq 0} (n + 1)! \|h_n\|^2_{L^2(Z^{n+1})} < \infty. \tag{10.5}$$

If u belongs to Dom δ, then the random variable $\delta(u)$ is defined as

$$\delta(u) = \sum_{n \geq 0} I_{n+1}(\tilde{h}_n),$$

where $\tilde{h}_n$ stands for the symmetrization of h as a function in $n + 1$ variables.

For instance, if $u(z) = h(z)$ is a deterministic function in $L^2(Z)$ then $\delta(u) = I_1(h)$. If $u(z) = I_1(h(\cdot, z))$, with $h \in L^2_s(Z^2)$, then $\delta(u) = I_2(h)$.

The following result characterizes δ as the adjoint of D.

Proposition 10.2.2 *If $u \in$ Dom δ, then $\delta(u)$ is the unique element of $L^2(\Omega)$ such that, for all $F \in \mathbb{D}^{1,2}$,*

$$\mathrm{E}(\langle DF, u \rangle_{L^2(Z)}) = \mathrm{E}(F\delta(u)).$$

Conversely, if u is a stochastic process in $L^2(\Omega \times Z)$ such that, for some $G \in L^2(\Omega)$ and for all $F \in \mathbb{D}^{1,2}$,

$$\mathrm{E}(\langle DF, u \rangle_{L^2(Z)}) = \mathrm{E}(FG),$$

then u belongs to Dom δ and $\delta(u) = G$.

Proof Let $F = \sum_{n \geq 0} I_n(g_n)$ and $u(z) = \sum_{n \geq 0} I_n(h_n(\cdot, z))$, where $g_n \in L^2_s(Z^n)$

and, for each $z \in Z$, $h_n(z, \cdot) \in L^2_s(Z^n)$. Then

$$
\begin{aligned}
E(\langle DF, u \rangle_{L^2(Z)}) &= E\left(\int_Z \sum_{n \geq 1} n I_{n-1}(g_n(\cdot, z)) \sum_{n \geq 0} I_n(h_n(\cdot, z)) \, m(dz) \right) \\
&= \int_Z E\left(\sum_{n \geq 0} (n+1) I_n(g_{n+1}(\cdot, z)) I_n(h_n(\cdot, z)) \right) m(dz) \\
&= \int_Z \sum_{n \geq 0} (n+1) n! \langle g_{n+1}(\cdot, z), h_n(\cdot, z) \rangle_{L^2(Z^n)} \, m(dz) \\
&= \sum_{n \geq 0} (n+1)! \langle g_{n+1}, h_n \rangle_{L^2(Z^{n+1})} = \sum_{n \geq 1} n! \langle g_n, \tilde{h}_{n-1} \rangle_{L^2(Z^n)}.
\end{aligned}
$$

Moreover, if $u \in \mathrm{Dom}\,\delta$ then

$$
\begin{aligned}
E(F\delta(u)) &= E\left(\sum_{n \geq 0} I_n(g_n) \sum_{n \geq 0} I_{n+1}(\tilde{h}_n) \right) = E\left(\sum_{n \geq 0} I_{n+1}(g_{n+1}) I_{n+1}(\tilde{h}_n) \right) \\
&= \sum_{n \geq 1} n! \langle g_n, \tilde{h}_{n-1} \rangle_{L^2(Z^n)}.
\end{aligned}
$$

Finally, uniqueness follows since $\mathbb{D}^{1,2}$ is dense in $L^2(\Omega)$. The second part of the proposition can be proved by similar arguments. $\square$

Proposition 10.2.2 implies that δ is the adjoint operator of D and it is clear that $\mathrm{Dom}\,\delta$ is dense in $L^2(\Omega \times Z)$ and that δ is a closed operator.

We next introduce the space of differentiable processes.

Definition 10.2.3   The space $\mathbb{L}^{1,2}$ is the set of processes u in $L^2(\Omega \times Z)$ such that $u(z)$ belongs to $\mathbb{D}^{1,2}$ for almost all z and there exists a measurable version of the two-parameter process Du such that

$$
E\left(\|Du\|^2_{L^2(Z^2)} \right) < \infty. \tag{10.6}
$$

The next result shows that $\mathbb{L}^{1,2}$ is included in $\mathrm{Dom}\,\delta$.

Lemma 10.2.4   *We have that $\mathbb{L}^{1,2} \subset \mathrm{Dom}\,\delta$.*

Proof   Let u be a stochastic process in $\mathbb{L}^{1,2}$ with representation

$$
u(z) = \sum_{n \geq 0} I_n(h_n(\cdot, z)),
$$

where, for each $z \in Z$, the function $h_n(\cdot, z)$ belongs to $L^2_s(Z^n)$. Observe that

(10.6) can be written as

$$E\left(\|Du\|^2_{L^2(Z^2)}\right) = E\left(\int_{Z^2} (D_\alpha u(z))^2 \, m(d\alpha)m(dz)\right)$$

$$= E\left(\int_{Z^2} \left(\sum_{n\geq 1} nI_{n-1}(h_n(\cdot, z, \alpha))\right)^2 m(d\alpha)m(dz)\right)$$

$$= \int_{Z^2} \sum_{n\geq 1} n^2(n-1)! \|h_n(\cdot, z, \alpha)\|^2_{L^2(Z^{n-1})} \, m(d\alpha)m(dz)$$

$$= \sum_{n\geq 1} (n+1)^2 n! \|h_n\|^2_{L^2(Z^{n+1})} < \infty,$$

which implies (10.5), and thus u belongs to Dom δ. □

We next compute the Malliavin derivative of the divergence operator.

Proposition 10.2.5 *Consider a stochastic process u in $L^{1,2}$ such that, for almost all $z \in Z$, the process $D_z u$ is in Dom δ and such that there exists a version of the process $(\delta(D_z u(\cdot)))_{z\in Z}$ that belongs to $L^2(\Omega \times Z)$. Then $\delta(u)$ belongs to $\mathbb{D}^{1,2}$ and*

$$D_z(\delta(u)) = u(z) + \delta(D_z u) \tag{10.7}$$

for almost all $z \in Z$.

Proof Suppose that u has the representation (10.4). Then

$$D_z(\delta(u)) = \sum_{n=0}^{\infty} (n+1)I_n(\tilde{h}_n(\cdot, z))$$

$$= u(z) + \sum_{n=0}^{\infty} I_n\left(\sum_{i=1}^{n} h_n(z_i, z_1, \ldots, z_{i-1}, z, z_{i+1}, \ldots, z_n)\right)$$

$$= u(z) + \sum_{n=0}^{\infty} nI_n(g_n(\cdot, z, \cdot)),$$

where $g_n(\cdot, z, \cdot)$ is the symmetrization of the function

$$(z_1, \ldots, z_n) \to h_n(z_1, \ldots, z_{n-1}, z, z_n).$$

Moreover,

$$\delta(D_z u) = \delta\left(\sum_{n=1}^{\infty} nI_{n-1}(h_n(\cdot, z, \cdot))\right) = \sum_{n=0}^{\infty} nI_n(g_n(\cdot, z, \cdot)),$$

which shows the desired result. □

The divergence operator δ satisfies the following product rule.

Proposition 10.2.6 *Let $F \in \mathbb{D}^{1,2}$ and $u \in \text{Dom}\,\delta$ be such that the product uDF belongs to* $\text{Dom}\,\delta$ *and the right-hand side of (10.8) below belongs to $L^2(\Omega)$. Then $Fu \in \text{Dom}\,\delta$ and*

$$\delta(Fu) = F\delta(u) - \langle DF, u \rangle_{L^2(Z)} - \delta(uDF). \tag{10.8}$$

Proof Let G be a bounded random variable in $\mathbb{D}^{1,2}$. Then, using Lemma 10.1.7 and Proposition 10.2.2, we obtain

$$\begin{aligned} E(\langle Fu, DG \rangle_{L^2(Z)}) &= E(\langle u, D(FG) - GDF - DFDG \rangle_{L^2(Z)}) \\ &= E(FG\delta(u)) - E(G\langle DF, u \rangle_{L^2(Z)}) - E(G\delta(uDF)) \\ &= E(G(F\delta(u) - \langle DF, u \rangle_{L^2(Z)} - \delta(uDF)). \end{aligned}$$

Finally, the result follows from Proposition 10.2.2.   □

We next consider the Poisson random measure N associated with a Lévy process. The next result shows that the divergence operator coincides with the stochastic integral with respect to the compensated measure $\hat{N}$ when it acts on predictable and square integrable processes.

Theorem 10.2.7 *Let u be a process in $L^2(\mathcal{P})$. Then u belongs to $\text{Dom}\,\delta$ and*

$$\delta(u) = \int_{\mathbb{R}_+} \int_{\mathbb{R}_0} u(t, z)\hat{N}(dt, dz).$$

Proof First assume that u is an elementary and predictable process of the form (9.9), where the $F_{i,j}$ belong to $\mathbb{D}^{1,2}$. We plan to apply Proposition 10.2.2 to the random variables $F_{i,j}$ and the processes $u_{i,j}(t, z) = \mathbf{1}_{(t_i, t_{i+1}]}(t)\mathbf{1}_{A_j}(z)$. Clearly $u_{i,j}$ belongs to $\text{Dom}\,\delta$. Furthermore, since the $F_{i,j}$ are $\mathcal{F}_{t_i}$-measurable for all $t > t_i$ and $\omega \in \Omega$, we have

$$D_{t,z}F_i(\omega) = F_{i,j}(\omega + \delta_{(t,z)}) - F_{i,j}(\omega) = 0,$$

which implies that $D_{t,z}F_{i,j}\mathbf{1}_{(t_i, t_{i+1}]}(t)\mathbf{1}_{A_j}(z) = 0$. Thus, by Proposition 10.2.6, we obtain

$$\begin{aligned} \delta(u) &= \sum_{j=1}^{m} \sum_{i=0}^{n-1} \Big(F_{i,j}\hat{N}((t_i, t_{i+1}], A_j) \\ &\qquad - \int_{t_i}^{t_{i+1}} \int_{A_j} D_{t,z}F_{i,j}\, \nu(dz)dt - \delta(DF_{i,j}\mathbf{1}_{(t_i, t_{i+1}]\times A_j}) \Big) \\ &= \sum_{j=1}^{m} \sum_{i=0}^{n-1} F_{i,j}\hat{N}((t_i, t_{i+1}], A_j), \end{aligned} \tag{10.9}$$

which coincides with the integral of u with respect to $\hat{N}$ (see (9.10)).

Next, assume that u is of the form (9.9), where the $F_{i,j}$ belong to $\mathcal{F}_i$ and are bounded. Then, since $\mathbb{D}^{1,2}$ is dense in $L^2(\Omega)$, there exists a sequence $F_{i,j}^k$ in $\mathbb{D}^{1,2}$ that converges to $F_{i,j}$ in $L^2(\Omega)$. The corresponding sequence u_k converges to u in $L^2(\Omega \times \mathbb{R}_+ \times \mathbb{R}_0)$ and (10.9) holds true for this sequence. Finally, since δ is a closed operator, we obtain that u belongs to Dom δ and that $\delta(u)$ is the limit in $L^2(\Omega)$ of $\delta(u_k)$, which also coincides with the integral of u with respect to $\hat{N}$.

Finally, the general case follows since the set $\mathcal{E}$ is dense in $L^2(\mathcal{P})$ and δ is a closed operator. $\qquad\qquad\qquad\qquad\qquad\qquad\qquad\qquad\qquad\qquad\square$

10.3 Ornstein–Uhlenbeck Semigroup

As in the Wiener case, we can define the Ornstein–Uhlenbeck semigroup and its generator. Let $F \in L^2(\Omega)$ be a random variable with chaos decomposition $F = \sum_{n=0}^{\infty} I_n(h_n)$, $h_n \in L_s^2(Z^n)$. Then the Ornstein–Uhlenbeck semigroup is defined as

$$T_t(F) = \sum_{n=1}^{\infty} e^{-nt} I_n(h_n).$$

We denote by L the generator of the Ornstein–Uhlenbeck semigroup. The domain of L, denoted Dom L, is given by those $F \in L^2(\Omega)$ such that

$$\sum_{n=1}^{\infty} n^2 n! \|h_n\|_{L^2(Z^n)}^2 < \infty.$$

If $F \in$ Dom L, the random variable LF is given by

$$LF = -\sum_{n=1}^{\infty} n I_n(h_n).$$

Note that $\mathrm{E}(LF) = 0$.

We also consider the pseudo-inverse L^{-1} of L. This is a bounded operator on the space of centered random variables in $L^2(\Omega)$. For every $F = \sum_{n=1}^{\infty} I_n(h_n)$, we have

$$LF = -\sum_{n=1}^{\infty} \frac{1}{n} I_n(h_n).$$

The relationship between the operators D, δ, and L follows easily, as in the Wiener case (Proposition 5.2.1), on applying the duality relationship (Proposition 10.2.2) and chaos expansions.

Proposition 10.3.1   *$F \in \text{Dom } L$ if and only if $F \in \mathbb{D}^{1,2}$ and $DF \in \text{Dom } \delta$. Moreover, in this case,*

$$\delta DF = -LF.$$

10.4 Clark–Ocone Formula

Consider the Poisson random measure N associated with a Lévy process in the filtered probability space $(\Omega, \mathcal{F}, (\mathcal{F}_t)_{t \geq 0}, P)$. We next present the jump version of the Clark–Ocone formula.

Theorem 10.4.1   *Let F be an $\mathcal{F}_T$-measurable random variable in $\mathbb{D}^{1,2}$. Then F admits the following representation:*

$$F = E(F) + \int_0^T \int_{\mathbb{R}_0} E(D_{t,z}F|\mathcal{F}_t)\hat{N}(dt, dz), \qquad (10.10)$$

where we have chosen a predictable version of the conditional expectation process $E(D_{t,z}F|\mathcal{F}_t)$.

Proof   By Theorem 9.6.1, because F is $\mathcal{F}_T$-measurable and belongs to $L^2(\Omega)$, it admits the representation

$$F = E(F) + \int_0^T \int_{\mathbb{R}_0} u(t, z)\hat{N}(dt, dz),$$

for some predictable process u in $L^2(\mathcal{P})$.

We now consider another predictable stochastic process $\tilde{u}$ in $L^2(\mathcal{P})$. The isometry property implies that on the one hand

$$E(\delta(\tilde{u})F) = \int_0^T \int_{\mathbb{R}_0} E\left(\tilde{u}(t, z)u(t, z)\right)\nu(dz)dt. \qquad (10.11)$$

On the other hand, by the duality relationship, we obtain

$$E(\delta(\tilde{u})F) = E\left(\int_0^T \int_{\mathbb{R}_0} \tilde{u}(t, z)D_{t,z}F\nu(dz)dt\right)$$

$$= \int_0^T \int_{\mathbb{R}_0} E\left(\tilde{u}(t, z)E(D_{t,z}F|\mathcal{F}_t)\right)\nu(dz)dt. \qquad (10.12)$$

Finally, (10.11) and (10.12) imply the desired result.   □

Example 10.4.2   Let $F = u_T^3$, where $u_T = \int_0^T \int_{\mathbb{R}_0} z\,\hat{N}(dt, dz)$. Then $D_{t,z}u_T = z\mathbf{1}_{[0,T]}(t)$. Moreover, by Exercise 10.1,

$$D_{t,z}F = (u_T + z\mathbf{1}_{[0,T]}(t))^3 - u_T^3 = \mathbf{1}_{[0,T]}(t)(z^3 + 3zu_T^2 + 3z^2u_T).$$

Therefore

$$E(D_{t,z}F|\mathcal{F}_t) = \mathbf{1}_{[0,T]}(t)\left(z^3 + 3z\left((T-t)\int_{\mathbb{R}_0} z^2 \nu(dz) + u_t^2\right) + 3z^2 u_t\right)$$

and

$$F = E(F) + \int_0^T \int_{\mathbb{R}_0} \left(z^3 + 3z\left((T-t)\int_{\mathbb{R}_0} z^2 \nu(dz) + u_t^2\right) + 3u_t z^2\right) \hat{N}(dt, dz).$$

10.5 Stein's Method for Poisson Functionals

In this section we combine Stein's method, introduced in Section 8.1, with the Malliavin calculus for Poisson random measures developed above, following Peccati *et al.* (2010).

Let M be a Poisson random measure on a complete separable metric space $(Z, \mathcal{Z}, m)$, where m is a σ-finite atomless measure. With the notation of Section 8.1, we have the following analogs of Theorem 8.2.1 and (8.15) in the Poisson case.

Theorem 10.5.1 *Let $F \in \mathbb{D}^{1,2}$ be such that $E(F) = 0$. Let X be an $N(0,1)$ random variable. Let $\mathcal{H}$ be a separating class of functions such that $E(|h(X)|) < \infty$ and $E(|h(F)|) < \infty$ for every $h \in \mathcal{H}$. Then*

$$d_{\mathcal{H}}(F, X) \leq \sup_{h \in \mathcal{H}} \|f_h'\|_\infty E\left(\left|1 - \langle DF, -DL^{-1}F\rangle_{L^2(Z)}\right|\right)$$
$$+ \frac{1}{2} \sup_{h \in \mathcal{H}} \|f_h''\|_\infty \int_Z E\left(|D_z F|^2 |D_z L^{-1}F|\right) m(dz).$$

Proof By Proposition 8.1.5, it suffices to bound

$$\sup_{h \in \mathcal{H}} |E(f_h'(F) - F f_h(F))|.$$

Using Theorem 10.1.4 twice, together with a second-order Taylor expansion, we get that, for all $\omega \in \Omega$ and $z \in Z$,

$$D_z f_h(F)(\omega) = f_h(F(\varphi_z(\omega))) - f_h(F(\omega))$$
$$= f_h'(F(\omega))(F(\varphi_z(\omega)) - F(\omega)) + R(F(\varphi_z(\omega)) - F(\omega))^2$$
$$= f_h'(F(\omega))D_z F(\omega) + R(D_z F(\omega))^2, \qquad (10.13)$$

where $R := R(\omega, z)$ satisfies $|R| \leq \frac{1}{2} \sup_{h \in \mathcal{H}} \|f_h''\|_\infty$.

Now, Propositions 10.2.2 and 10.3.1 with $u = DL^{-1}F$ and $G = f_h(F)$

yield

$$E(F f_h(F)) = E\left(LL^{-1} F f_h(F)\right) = -E\left(\delta(DL^{-1}F) f_h(F)\right)$$
$$= E\left(\langle D f_h(F), -DL^{-1}F\rangle_{L^2(Z)}\right).$$

According to (10.13), we have

$$E\left(\langle D f_h(F), -DL^{-1}F\rangle_{L^2(Z)}\right) = E\left(f_h'(F)\langle DF, -DL^{-1}F\rangle_{L^2(Z)}\right)$$
$$+ E\left(R\langle (DF)^2, -DL^{-1}F\rangle_{L^2(Z)}\right).$$

It follows that

$$|E(f_h'(F) - F f_h(F))| \le \left|E\left(f_h'(F)(1 - \langle DF, -DL^{-1}F\rangle_{L^2(Z)})\right)\right|$$
$$+ \left|E\left(R\langle (DF)^2, -DL^{-1}F\rangle_{L^2(Z)}\right)\right|.$$

The result follows immediately. □

We are interested in the case where the class $\mathcal{H} = \text{Lip}(1)$, which corresponds to Wasserstein's distance. For any $h \in \text{Lip}(1)$, by Proposition 8.1.8, $\|f_h'\|_\infty \le \sqrt{2/\pi} \le 1$ and, by Stein (1986, Lemma 3, eq. (47), p. 25), $\|f_h''\|_\infty \le 2$. Therefore, for any $F \in \mathbb{D}^{1,2}$ such that $E(F) = 0$, Theorem 10.5.1 yields

$$d_W(F, X) \le E\left(|1 - \langle DF, -DL^{-1}F\rangle_{L^2(Z)}|\right) + \int_Z E\left(|D_z F|^2 |D_z L^{-1} F|\right) m(dz).$$
$$(10.14)$$

10.6 Normal Approximation on a Fixed Chaos

Suppose that $F = I_q(f) \in \mathcal{H}_q, q \ge 1$, and $f \in L_s^2(Z^q)$. By the definition of L^{-1}, we can write

$$\langle DF, -DL^{-1}F\rangle_{L^2(Z)} = \frac{1}{q}\|DF\|_{L^2(Z)}^2$$

and

$$\int_Z E\left(|D_z F|^2 |D_z L^{-1} F|\right) m(dz) = \frac{1}{q}\int_Z E\left(|D_z F|^3\right) m(dz).$$

As a consequence, from (10.14) we obtain

$$d_W(F, X) \le E\left(|1 - q^{-1}\|DF\|_{L^2(Z)}^2|\right) + \frac{1}{q}\int_Z E\left(|D_z F|^3\right) m(dz). \quad (10.15)$$

The Case $q = 1$

We now fix $h \in L^2(Z)$. Then $I_1(h) = \hat{M}(h)$ belongs to $\mathbb{D}^{1,2}$ and $D\hat{M}(h) = h$. Moreover, $-L^{-1}\hat{M}(h) = \hat{M}(h)$. Therefore, (10.15) yields the following result.

Corollary 10.6.1 *Let $h \in L^2(Z)$ and let X be an $N(0, 1)$ random variable. Then the following bound holds:*

$$d_W(\hat{M}(h), X) \leq \left|1 - \|h\|^2_{L^2(Z)}\right| + \int_Z |h(z)|^3 m(dz). \quad (10.16)$$

As a consequence, if $m(Z) = \infty$ and the sequence $(h_k)_{k\geq 1} \subset L^2(Z) \cap L^3(Z)$ satisfies, as $k \to \infty$,

$$\|h_k\|_{L^2(Z)} \to 1 \quad \text{and} \quad \|h_k\|_{L^3(Z)} \to 0$$

then we have the central limit theorem

$$\hat{M}(h_k) \overset{\mathcal{L}}{\longrightarrow} X$$

and inequality (10.16) provides an explicit upper bound for the Wasserstein distance.

Example 10.6.2 Consider a Poisson random measure $\hat{M}$ on $Z = \mathbb{R}_+$, with measure m equal to the Lebesgue measure. Then the random variable $k^{-1/2}\hat{M}([0, k]) = \hat{M}(h_k)$, where $h_k = k^{-1/2}\mathbf{1}_{[0,k]}$, is an element of the first Wiener chaos associated with $\hat{M}$. Since the random variables $\hat{M}((i - 1, i])$, $i = 1, \ldots, k$, are iid centered Poisson with unit variance, a standard application of the central limit theorem yields that, as $k \to \infty$, $\hat{M}(h_k) \overset{\mathcal{L}}{\to} X$ where X is an $N(0, 1)$ random variable. Moreover, $\hat{M}(h_k)$ belongs to $\mathbb{D}^{1,2}$ for every k, and $D\hat{M}(h_k) = h_k$. Since

$$\|h_k\|_{L^2(Z)} = 1 \quad \text{and} \quad \int_Z |h_k(z)|^3 m(dz) = \frac{1}{k^{1/2}},$$

one deduces from Corollary 10.6.1 that

$$d_W(\hat{M}(h), X) \leq \frac{1}{k^{1/2}},$$

which is consistent with the usual Berry–Esséen estimates.

The Case $q \geq 2$

Let $f \in L^2_s(Z^q)$. We consider an operator G^q_p that transforms a function of q variables into a function of p variables, as follows. When $p = 0$, we set

$$G^q_0 f = q! \|f\|^2_{L^2(Z^q)}$$

and, for every $p = 1, \ldots, 2q$, we set

$$(G_p^q f)(z_1, \ldots, z_p) = \sum_{r=0}^{q} \sum_{\ell=0}^{r} \mathbf{1}_{\{2q-r-\ell=p\}} r! \binom{q}{r}^2 \binom{r}{\ell} \left(f \widetilde{\star}_r^{\ell} f \right)(z_1, \ldots, z_p).$$

This operator allows us to give a more compact representation of the product formula of two multiple integrals (9.22) when $f = g$. In fact, if $f \star_r^{\ell} f \in L^2(Z^{2q-r-\ell})$ for all $r = 0, \ldots, q$ and $\ell = 1, \ldots, r$ then one has that

$$I_q(f)^2 = \sum_{p=0}^{2q} I_p(G_p^q f).$$

From now on, we assume that every contraction of the type $f \star_r^{\ell} f$ is well defined and finite for every $r = 0, \ldots, q$ and $\ell = 1, \ldots, r$.

When $m(Z) = \infty$, we consider the following technical condition, which is needed to justify a Fubini argument: for every $p = 1, \ldots, 2(q-1)$,

$$\int_Z \left(\int_{Z^p} (G_p^{q-1} f(z, \cdot))^2 dm^p \right)^{1/2} m(dz) < \infty. \tag{10.17}$$

Finally, we set

$$\widehat{G}_p^q f(\cdot) = \int_Z G_p^{q-1} f(z, \cdot) m(dz).$$

We have the following Wasserstein bounds on a fixed chaos, proved by Peccati *et al.* (2010, Theorem 4.2).

Theorem 10.6.3 *Let X be an $N(0, 1)$ random variable. Fix $q \geq 2$, and let $f \in L_s^2(Z^q)$ such that:*

(i) *whenever $m(Z) = \infty$, condition (10.17) is satisfied;*
(ii) *the kernel $f(z, \cdot) \star_r^{\ell} f(z, \cdot)$ belongs to $L^2(Z^{2(q-1)-r-\ell})$, for dm-almost-every $z \in Z$, $r = 1, \ldots, q-1$ and $\ell = 0, \ldots, r-1$.*

Then

$$d_W(I_q(f), X)$$

$$\leq \left(\left(1 - q! \|f\|_{L^2(Z^q)}^2 \right)^2 + q^2 \sum_{p=1}^{2(q-1)} p! \int_{Z^p} \left(\widehat{G}_p^q f \right)^2 dm^p \right)^{1/2}$$

$$+ q^2 \left((q-1)! \|f\|_{L^2(Z^q)}^2 \sum_{p=1}^{2(q-1)} p! \int_Z \left(\int_{Z^p} (G_p^{q-1} f(z, \cdot))^2 dm^p \right) m(dz) \right)^{1/2}.$$

$$\tag{10.18}$$

Moreover,

$$\left(\left(1 - q!\|f\|^2_{L^2(Z^q)}\right)^2 + q^2 \sum_{p=1}^{2(q-1)} p! \int_{Z^p} \left(\widehat{G^q_p f}\right)^2 dm^p\right)^{1/2}$$

$$\le \left|1 - q!\|f\|^2_{L^2(Z^q)}\right| + q \sum_{t=1}^{q} \sum_{s=1}^{t\wedge(q-1)} 1_{\{2\le t+s\le 2q-1\}}((2q - t - s)!)^{1/2}$$

$$\times (t - 1)! \binom{q-1}{t-1}^2 \binom{t-1}{s-1} \|f \star^s_t f\|_{L^2(Z^{2q-t-s})}.$$

(10.19)

Furthermore, if

$$f \star^q_{q-b} f \in L^2(Z^b) \quad \text{for all } b = 1, \ldots, q-1 \tag{10.20}$$

then

$$\left(\sum_{p=0}^{2(q-1)} p! \int_Z \left(\int_{Z^p} (G^{q-1}_p f(z, \cdot))^2 dm^p\right) m(dz)\right)^{1/2}$$

$$\le \sum_{b=1}^{q} \sum_{a=0}^{b-1} 1_{\{1\le a+b\le 2q-1\}}((a + b - 1)!)^{1/2}(q - a - 1)!$$

$$\times \binom{q-1}{q-a-1}^2 \binom{q-a-1}{q-b} \|f \star^a_b f\|_{L^2(Z^{2q-a-b})}.$$

(10.21)

Example 10.6.4 Consider a double integral $I_2(f)$, where $f \in L^2_s(Z^2)$ satisfies (10.17). Since in this case $G^1_1 f(z, \cdot)(x) = f^2(z, x)$ and $G^2_1 f(z, \cdot)(x, y) = f(z, x)f(z, y)$, we obtain that (10.17) holds if and only if

$$\int_Z \left(\int_Z f^4(z, x)m(dx)\right)^{1/2} m(dz) < \infty.$$

We assume the following conditions:

$$E(I_2(f)^2) = 1, \quad f \star^1_2 f \in L^2(Z), \quad \text{and} \quad f \in L^4(Z^2). \tag{10.22}$$

Then, assumptions (ii) and (10.20) of Theorem 10.6.3 are satisfied, since $f(z, \cdot) \star^0_1 f(z, \cdot)(x) = f^2(z, x)$, which is square integrable. Therefore, from (10.18)–(10.21), we obtain that the Wasserstein distance between the law of $I_2(f)$ and the law of an $N(0, 1)$ random variable X satisfies the following bound:

$$d_W(I_2(f), X) \le 2\left(\sqrt{2}\|f \star^1_1 f\|_{L^2(Z^2)} + \|f \star^1_2 f\|_{L^2(Z^2)}\right)$$

$$+ \sqrt{8}\left(\sqrt{2}\|f \star^0_1 f\|_{L^2(Z^3)} + \|f \star^0_2 f\|_{L^2(Z^2)} + \|f \star^1_2 f\|_{L^2(Z)}\right).$$

We now use the equalities

$$\|f \star_1^0 f\|_{L^2(Z^3)} = \|f \star_2^1 f\|_{L^2(Z)}$$

and

$$\|f \star_2^0 f\|_{L^2(Z^2)} = \|f\|_{L^4(Z^2)}^2,$$

which follow from a Fubini argument, to conclude that

$$dw(I_2(f), X) \le \sqrt{8}\|f \star_1^1 f\|_{L^2(Z^2)} + (2 + \sqrt{8}(1 + \sqrt{2}))\|f \star_2^1 f\|_{L^2(Z^2)}$$
$$+ \sqrt{8}\|f\|_{L^4(Z^2)}^2.$$

We thus have the following central limit theorem for multiple integrals of arbitrary order, due to Peccati *et al.* (2010, Theorem 5.1).

Theorem 10.6.5 *Suppose that $m(Z) = \infty$. Fix $q \ge 2$, and consider a sequence of multiple stochastic integrals of order q, $F_n = I_q(f_n) \in \mathcal{H}_q$, $n \ge 1$, such that*

$$\lim_{n \to \infty} E(F_n^2) = 1.$$

Assume moreover that the following conditions hold:

(i) f_n *satisfies* (10.17) *for any $n \ge 1$.*
(ii) $f_n \star_r^\ell f_n \in L^2(Z^{2q-r-\ell})$, *for any $r = 1, \ldots, q$ and $\ell = 0, \ldots, r \wedge (q-1)$, and $\|f_n \star_r^\ell f_n\|_{L^2(Z^{2q-r-\ell})} \to 0$ as $n \to \infty$.*
(iii) $f_n \in L^4(Z^q)$, *for any $n \ge 1$, and $\|f_n\|_{L^4(Z^q)} \to 0$ as $n \to \infty$.*

Then, as $n \to \infty$, $F_n \xrightarrow{\mathcal{L}} N(0, 1)$.

Proof Assumption (ii) implies that assumptions (ii) and (10.20) in the statement of Theorem 10.6.3 hold for any $f_n, n \ge 1$. Then relations (10.18)–(10.21) imply that $dw(F_n, X) \to 0$ as $n \to \infty$. Since convergence in the Wasserstein distance implies convergence in law, the conclusion holds. $\square$

Example 10.6.6 We consider a sequence of double integrals $(I_2(f_n))_{n \ge 1}$, where the f_n satisfy the same conditions as f of Example 10.6.4. Then, according to Theorem 10.6.5, the sufficient conditions in order that, as $n \to \infty$, we have

$$I_2(f_n) \xrightarrow{\mathcal{L}} N(0, 1) \tag{10.23}$$

are that $\|f_n\|_{L^4(Z^q)} \to 0$,

$$\|f_n \star_2^1 f_n\|_{L^2(Z)} \to 0, \quad \text{and} \quad \|f_n \star_1^1 f_n\|_{L^2(Z^2)} \to 0. \tag{10.24}$$

A counterpart of this last example is the following result, proved by Peccati and Taqqu (2008).

Proposition 10.6.7 *Consider a sequence of double integrals $(I_2(f_n))_{n\geq 1}$, where the f_n satisfy conditions (10.22) of Example 10.6.4. Assume also that as $n \to \infty$, $\|f_n\|_{L^4(Z^q)} \to 0$. Then,*

(i) *If $F_n \in L^4(\Omega)$ for every $n \geq 1$, a sufficient condition for (10.24) to hold is that*

$$E(F_n^4) \to 3. \tag{10.25}$$

(ii) *If the sequence F_n is uniformly integrable, conditions (10.23), (10.24), and (10.25) are equivalent.*

We end this section by presenting the analog of the fourth-moment theorem in the Poisson framework. A first proof of this result was given by Döbler and Peccati (2018) with the additional condition that the processes $\varphi(F)$, $F\varphi(F)$, $(\varphi(F))^4$, and $F^3\varphi(F)$ belong to $L^1(\Omega \times Z)$. Then, Döbler *et al.* (2018) were able to remove this condition and extend the theorem to the multidimensional case.

Theorem 10.6.8 *Fix $q \geq 1$ and let $F = I_q(f)$ be a multiple stochastic integral with $E(F^2) = \sigma^2 > 0$. Then, if X is an $N(0, \sigma^2)$ random variable,*

$$d_W(F, X) \leq \left(\frac{1}{\sigma} \sqrt{\frac{2}{\pi}} + \frac{4}{3\sigma} \right) \sqrt{E(F^4) - 3\sigma^4}.$$

Corollary 10.6.9 *For each $n \geq 1$, let $q_n \geq 1$ and let $F_n = I_{q_n}(f_n) \in \mathcal{H}_{q_n}$, $f_n \in L_s^2(Z^{q_n})$, be a sequence of multiple stochastic integrals satisfying*

$$\lim_{n\to\infty} E(F_n^2) = 1 \quad and \quad \lim_{n\to\infty} E(F_n^4) = 3.$$

Then, as $n \to \infty$, $F_n \xrightarrow{\mathcal{L}} N(0, 1)$.

We refer to Peccati and Reitzner (2016) for geometric applications, such as to random graphs, of the results in this section.

Exercises

10.1 Show by induction that if F belongs to $\mathbb{D}^{1,2}$ then

$$D(F^n) = (F + DF)^n - F^n.$$

10.2   Consider the process $Y = (Y_t)_{t\in[0,T]}$ given by

$$Y_t = \exp\left(\int_0^t \int_{\mathbb{R}_0} h_s z \hat{N}(ds, dz)\right.$$

$$\left. - \int_0^t \int_{\mathbb{R}_0} \left(e^{h_s z} - 1 - h_s z\right) \nu(dz) ds\right),$$

where $h \in L^2([0, T])$ is càdlàg. Compute the Malliavin derivative of Y_T using the chaos expansion obtained in Exercise 9.9.

10.3   Consider the process $X_t = \int_0^t \int_{\mathbb{R}_0} z \hat{N}(ds, dz)$, $t \in [0, T]$. Using the chaos expansion obtained in Exercise 9.10, compute the Malliavin derivative of the following random variables:

$$\int_0^T X_t dt, \quad X_T^3, \quad e^{X_T}, \quad \sin X_T.$$

10.4   Compute the Malliavin derivative of the random variables e^{X_T} and $\sin(X_T)$ in Exercise 10.3 using the chain rule (Proposition 10.1.8).

10.5   Using the Clark–Ocone formula, find integral representations for the random variables in Exercise 10.3.

11

Malliavin Calculus for Jump Processes II

In this chapter we define a derivative operator for cylindrical functionals of the Poisson random measure associated with a Lévy process. This differential operator is different from the one introduced in Chapter 10 via the chaos expansion. The Malliavin calculus for jump processes presented in this chapter follows ideas from Bichteler and Jacod (1983) in the case where $f(z) = \mathbf{1}_{\{|z|\leq 1\}}$ in (11.3) below (see also Bichteler *et al.*, 1987, for the multidimensional case), where f is the density of the Lévy measure. This approach was extended by Fournier (2000) for a general f as in (11.3). The derivative operator defined in this chapter satisfies the usual chain rule. Although we will not consider the adjoint operator, we derive the integration-by-parts formula (11.11), that allows us to obtain a criterion for the existence of the density of a random variable in the Wiener–Poisson framework. We apply this criterion to a diffusion with jumps.

11.1 Derivative Operator

Let $L = (L_t)_{t\geq 0}$ be a pure-jump Lévy process with characteristic triplet $(0, 0, \nu)$ defined on a complete probability space $(\Omega, \mathcal{F}, P)$ such that $\mathcal{F}$ is generated by L. Consider the Poisson random measure N associated with L, and its compensated random measure $\hat{N}$. We will make use of the notation

$$N(h) = \int_{\mathbb{R}_+} \int_{\mathbb{R}_0} h(t, z) N(dt, dz),$$

for $h \in L^1(\mathbb{R}_+ \times \mathbb{R}_0, \ell \times \nu)$. We denote by $C_0^{0,2}(\mathbb{R}_+ \times \mathbb{R}_0)$ the set of continuous functions $h \colon \mathbb{R}_+ \times \mathbb{R}_0 \to \mathbb{R}$ that have compact support and are twice differentiable on $\mathbb{R}_0$.

We consider the set $\mathcal{S}$ of cylindrical random variables of the form

$$F = \varphi(N(h_1), \ldots, N(h_m)), \tag{11.1}$$

where $\varphi \in C_0^2(\mathbb{R}^m)$ and $h_i \in C_0^{0,2}(\mathbb{R}_+ \times \mathbb{R}_0)$ for $1 \leq i \leq m$.

It is easy to show that the set $\mathcal{S}$ is dense in $L^2(\Omega)$.

Definition 11.1.1 The *Malliavin derivative* of a simple random variable F in $\mathcal{S}$ of the form (11.1) is defined as the two parameter process

$$D_{t,z}F = \sum_{k=1}^{m} \frac{\partial \varphi}{\partial x_k}(N(h_1), \ldots, N(h_m))\partial_z h_k(t, z), \quad (t, z) \in \mathbb{R}_+ \times \mathbb{R}_0. \quad (11.2)$$

In particular, $D(N(h)) = \partial_z h$.

Remark 11.1.2 Let $F \in \mathcal{S}$ be of the form (11.1). Then, there exist $T > 0$ and K a compact subset of $\mathbb{R}_0$ such that $[0, T] \times K$ contains the supports of all the h_i. Let $(T_i, Y_i)_{1 \leq i \leq M}$ denote the points of $[0, T] \times K$ that belong to the support of N with $T_1 < \cdots < T_M$ (if there are no such points, $M = 0$). Then M is a Poisson random variable with parameter $T\nu(K)$, and conditionally on M and $T_1, \ldots, T_M$, the random variables $(Y_i)_{1 \leq i \leq M}$ are independent and have distribution $\nu/\nu(K)$ over K.

From now on we assume the following hypothesis:

(H1) There exists a strictly positive function f in $C^1(\mathbb{R}_0)$ such that the Lévy measure satisfies

$$\nu(dz) = f(z)dz. \quad (11.3)$$

For any F in $\mathcal{S}$ of the form (11.1), we consider the operator given by

$$LF = \frac{1}{2} \sum_{k=1}^{m} \frac{\partial \varphi}{\partial x_k}(N(h_1), \ldots, N(h_m))N\left(\partial_z^2 h_k + \frac{f'}{f}\partial_z h_k\right)$$

$$+ \frac{1}{2} \sum_{k,j=1}^{m} \frac{\partial^2 \varphi}{\partial x_k \partial x_j}(N(h_1), \ldots, N(h_m))N\left(\partial_z h_k \partial_z h_j\right).$$

Consider the measure ν_N on $\mathbb{R}_+ \times \mathbb{R}_0$ given by

$$\nu_N(A) = E\left(\int_{\mathbb{R}_+} \int_{\mathbb{R}_0} \mathbf{1}_A(s, z)N(ds, dz)\right), \quad \text{for } A \in \mathcal{B}(\mathbb{R}_+ \times \mathbb{R}_0).$$

Given random elements $u, \tilde{u}$ in $L^2(\nu_N) := L^2(\mathbb{R}_+ \times \mathbb{R}_0, \nu_N)$, we define the random scalar product and norm

$$\langle u, \tilde{u} \rangle_N = \int_{\mathbb{R}_+} \int_{\mathbb{R}_0} u(s, z)\tilde{u}(s, z)N(ds, dz) \quad \text{and} \quad \|u\|_N^2 = \langle u, u \rangle_N.$$

Notice that $L^2(\mathcal{P}) \subset L^2(\nu_N)$ and, for any $u \in L^2(\mathcal{P})$, we have (Exercise 11.1)

$$E(\|u\|_N^2) = \int_{\mathbb{R}_+} \int_{\mathbb{R}_0} E(u^2(s, z))\nu(dz)ds. \quad (11.4)$$

The following result, whose proof is an immediate consequence of the definition of the operators D and L, shows the relationship between these operators (see Exercise 11.2).

Lemma 11.1.3 *For any $F, G \in S$, we have*

$$\langle DF, DG \rangle_N = L(FG) - FLG - GLF.$$

Let F be an element of S of the form (11.1). By Remark 11.1.2

$$F = \Phi((T_i, Y_i)_{1 \le i \le M}),$$

where

$$\Phi((t_i, y_i)_{1 \le i \le M}) = \varphi\left(\sum_{i=1}^{M} h_1(t_i, y_i), \ldots, \sum_{i=1}^{M} h_m(t_i, y_i) \right).$$

For any $1 \le i \le M$ and $\lambda \in \mathbb{R}$, define

$$F^{i,\lambda} = \Phi((T_1, Y_1), \ldots, (T_i, Y_i + \lambda), \ldots, (T_M, Y_M)). \tag{11.5}$$

The next lemma provides an alternative representation of LF (see Exercise 11.3).

Lemma 11.1.4 *Let $F \in S$ be of the form (11.1). Then*

$$LF = \tfrac{1}{2} \sum_{i=1}^{M} \frac{1}{f(Y_i)} \frac{d}{d\lambda} \left(f(Y_i + \lambda) \frac{dF^{i,\lambda}}{d\lambda} \right) \bigg|_{\lambda=0},$$

where $F^{i,\lambda}$ is given by (11.5).

As a consequence, if $F = 0$, conditioning on $T_1, \ldots, T_M$ and M, we have

$$\Phi((T_i, y_i)_{1 \le i \le M}) = 0$$

for all $y_1, \ldots, y_M \in \mathbb{R}_0$. This implies that $F^{i,\lambda} = 0$ and, therefore, $F = 0$ and $\|DF\|_N = 0$.

Remark 11.1.5 Assume that Ω is the canonical space (9.4) with $Z = \mathbb{R}_+ \times \mathbb{R}_0$. Then, for any F in S and $\omega \in \Omega$, if (t, z) is in the support of ω and $\lambda \in \mathbb{R}$ is such that $z + \lambda \ne 0$,

$$D_{t,z} F(\omega) = \frac{\partial}{\partial \lambda} F(\omega - \delta_{(t,z)} + \delta_{(t,z+\lambda)}) \bigg|_{\lambda=0}.$$

The next integration-by-parts formula is due to Bichteler *et al.* (1987, Proposition 9.3d).

Lemma 11.1.6 *For any $F, G \in S$, it holds that*

$$E(FLG) = E(GLF) = -\tfrac{1}{2}E(\langle DF, DG \rangle_N).$$

Proof Let $F = \varphi(N(h_1), \ldots, N(h_m))$ and $G = \psi(N(g_1), \ldots, N(g_q))$. Let $T > 0$, and let K be a compact subset of $\mathbb{R}_0$ such that $[0, T] \times K$ contains the supports of all h_i and g_i, as in Remark 11.1.2. Let $(T_i, Y_i)_{1 \leq i \leq M}$ denote the points of $[0, T] \times K$ that belong to the support of N with $T_1 < \cdots < T_M$. We denote by $\mathcal{G}$ the σ-field generated by M and $T_1, \ldots, T_M$.

We will show that

$$E\left(FLG + \tfrac{1}{2}\langle DF, DG \rangle_N \,\middle|\, \mathcal{G}\right) = 0,$$

which implies that $E(FLG) = -\tfrac{1}{2}E(\langle DF, DG \rangle_N)$. We can deduce the other equality by the symmetry between F and G.

Let M and the T_i be fixed, and set $h_i^j(z) = h_i(T_j, z)$ and $g_i^j(z) = g_i(T_j, z)$. Then, $FLG + \tfrac{1}{2}\langle DF, DG \rangle_N = \Psi(Y_1, \ldots, Y_M)$, where

$$\Psi(z_1, \ldots, z_M)$$

$$= \tfrac{1}{2}\varphi\left(\sum_{\ell=1}^{M} h_1^\ell(z_\ell), \ldots, \sum_{\ell=1}^{M} h_m^\ell(z_\ell)\right)$$

$$\times \left(\sum_{k=1}^{q} \frac{\partial\psi}{\partial x_k}\left(\sum_{\ell=1}^{M} g_1^\ell(z_\ell), \ldots, \sum_{\ell=1}^{M} g_q^\ell(z_\ell)\right) \sum_{\ell=1}^{M}\left(\partial_z^2 g_k^\ell(z_\ell) + \frac{f'(z_\ell)}{f(z_\ell)}\partial_z g_k^\ell(z_\ell)\right)\right.$$

$$+ \sum_{k,j=1}^{q} \frac{\partial^2\psi}{\partial x_k \partial x_j}\left(\sum_{\ell=1}^{M} g_1^\ell(z_\ell), \ldots, \sum_{\ell=1}^{M} g_q^\ell(z_\ell)\right) \sum_{\ell=1}^{M} \partial_z g_k^\ell(z_\ell)\partial_z g_j^\ell(z_\ell)\right)$$

$$+ \tfrac{1}{2}\sum_{k,j=1}^{m} \frac{\partial\varphi}{\partial x_k}\left(\sum_{\ell=1}^{M} h_1^\ell(z_\ell), \ldots, \sum_{\ell=1}^{M} h_m^\ell(z_\ell)\right) \frac{\partial\psi}{\partial x_j}\left(\sum_{\ell=1}^{M} g_1^\ell(z_\ell), \ldots, \sum_{\ell=1}^{M} g_q^\ell(z_\ell)\right)$$

$$\times \sum_{\ell=1}^{M} \partial_z h_k^\ell(z_\ell)\partial_z g_j^\ell(z_\ell).$$

Then, if μ_K denotes the restriction of ν to K, it suffices to show that

$$\int_{K^M} \Psi(z_1, \ldots, z_M)\mu_K(dz_1) \cdots \mu_K(dz_M) = 0. \tag{11.6}$$

Let $1 \leq \ell \leq M$ and $\hat{z}_\ell = \{z_j\}_{1 \leq j \leq M, j \neq \ell}$. Then, one can easily check that

$$\Psi(z_1, \ldots, z_M) = \tfrac{1}{2}\sum_{\ell=1}^{M} \frac{1}{f(z_\ell)}\partial_{z_\ell} Q_{\hat{z}_\ell}^\ell(z_\ell),$$

where

$$Q_{\hat{z}_\ell}^\ell(z_\ell) = \varphi\left(\sum_{i=1}^M h_1^i(z_i), \ldots, \sum_{i=1}^M h_m^i(z_i)\right)$$

$$\times \sum_{k=1}^q \frac{\partial\psi}{\partial x_k}\left(\sum_{i=1}^M g_1^i(z_i), \ldots, \sum_{i=1}^M g_q^i(z_i)\right) f(z_\ell)\partial_z g_k^\ell(z_\ell).$$

Since $Q_{\hat{z}_\ell}^\ell = 0$ on the boundary of K, we deduce that, for all ℓ and $\hat{z}_\ell$,

$$\int_K \frac{1}{f(z)}\partial_z Q_{\hat{z}_\ell}^\ell(z)\mu_K(dz) = \int_K \partial_z Q_{\hat{z}_\ell}^\ell(z)dz = 0,$$

from which we conclude (11.6). □

11.2 Sobolev Spaces

As a consequence of Lemma 11.1.6, we will show that the operator D is closable.

Lemma 11.2.1 *Let F_k be a sequence of random variables in S that converges to zero in $L^2(\Omega)$. Assume that there exists a process u in $L^2(\nu_N)$ such that $E(\|DF_k - u\|_N^2)$ converges to zero. Then $E(\|u\|_N^2) = 0$.*

Proof Let $F \in S$. Then,

$$E(\langle u, DF\rangle_N) = \lim_{k\to\infty} E(\langle DF_k, DF\rangle_N).$$

Thanks to Lemma 11.1.6, $E(\langle DF_k, DF\rangle_N) = -2E(F_kLF)$. Since F_k converges to zero in $L^2(\Omega)$, we deduce that $E(\langle u, DF\rangle_N) = 0$. We then apply this with $F = F_k$ and let k go to ∞. □

We denote by $\mathbb{D}^{1,N}$ the closure of S with respect to the seminorm

$$\|F\|_{1,N} = \left(E(|F|^2) + E(\|DF\|_N^2)\right)^{1/2}.$$

Notice that to show that a random variable F in $L^2(\Omega)$ belongs to $\mathbb{D}^{1,N}$ and that, for some process u in $L^2(\nu_N)$, $DF = u$ it suffices to find a sequence F_k in S such that

$$E(|F - F_k|^2) + E(\|u - DF_k\|_N^2)$$

converges to zero as k goes to infinity. Observe also that $\mathbb{D}^{1,N}$ is a Hilbert space with scalar product

$$\langle F, G\rangle_{1,N} = E(FG) + E(\langle DF, DG\rangle_N).$$

The next result is the chain rule for the Malliavin derivative in the Poisson framework, which can be proved as in the Brownian motion case (see Proposition 3.3.2).

Proposition 11.2.2   *Let φ be a function in $C^1(\mathbb{R})$ with bounded derivative, and let F be a random variable in $\mathbb{D}^{1,N}$. Then, $\varphi(F)$ belongs to $\mathbb{D}^{1,N}$ and*

$$D(\varphi(F)) = \varphi'(F)DF.$$

Denote by $C_b^{0,1}(\mathbb{R}_+ \times \mathbb{R}_0)$ the set of continuous functions $h \colon \mathbb{R}_+ \times \mathbb{R}_0 \to \mathbb{R}$ that are differentiable on $\mathbb{R}_0$ with

$$\int_{\mathbb{R}_+} \int_{\mathbb{R}_0} \left(h^2(t,z) + (\partial_z h(t,z))^2 \right) \nu(dz) dt < \infty.$$

Proposition 11.2.3   *If $h \in C_b^{0,1}(\mathbb{R}_+ \times \mathbb{R}_0)$ then $N(h)$ and $\hat{N}(h)$ belong to $\mathbb{D}^{1,N}$ and*

$$DN(h) = D\hat{N}(h) = \partial_z h.$$

Proof   The result is clearly true if $h \in C_0^{0,2}(\mathbb{R}_+ \times \mathbb{R}_0)$. The general case will follow from an approximation argument. For any $\epsilon > 0$, consider a nonnegative function in $C^2(\mathbb{R}_0)$ such that

$$K_\epsilon(z) = \begin{cases} 1 & \text{if } |z| < 1/\epsilon, \\ 0 & \text{if } |z| > 1/\epsilon + 2, \end{cases}$$

$|K_\epsilon(z)| \leq 1$, and $|K_\epsilon'(z)| \leq 1$. Consider also a $C(\mathbb{R}_+)$ nonegative function such that

$$M_\epsilon(t) = \begin{cases} 1 & \text{if } t \leq 1/\epsilon, \\ 0 & \text{if } t \geq 1 + 1/\epsilon, \end{cases}$$

and $|M_\epsilon(t)| \leq 1$. Set

$$h_\epsilon(t,z) = (h * \varphi_\epsilon)(t,z) K_\epsilon(z) M_\epsilon(t),$$

where $h \in C_b^{0,1}(\mathbb{R}_+ \times \mathbb{R}_0)$ and φ_ϵ is an approximation of the identity on $\mathbb{R}_0$. Then, for all $\epsilon > 0$, h_ϵ belongs to $C_0^{0,2}(\mathbb{R}_+ \times \mathbb{R}_0)$. Therefore, $N(h_\epsilon)$ and $\hat{N}(h_\epsilon)$ belong to $\mathbb{D}^{1,N}$, and

$$DN(h_\epsilon) = D\hat{N}(h_\epsilon) = \partial_z h_\epsilon = (\partial_z h * \varphi_\epsilon) K_\epsilon M_\epsilon + (h * \varphi_\epsilon) K_\epsilon' M_\epsilon.$$

Finally, one can easily check that, as ϵ tends to 0, h_ϵ converges to h in $L^1(\ell \times \nu)$ and $L^2(\ell \times \nu)$, $N(h_\epsilon)$ and $\hat{N}(h_\epsilon)$ converge to $N(h)$ and $\hat{N}(h)$ in $L^2(\Omega)$, respectively, and $\mathrm{E}(\|\partial_z h - \partial_z h_\epsilon\|_N^2)$ converges to 0. This concludes the proof. $\square$

We now introduce the space of adapted differentiable random processes in the Malliavin sense. Denote by $\mathcal{E}_a$ the space of elementary and predictable processes of the form

$$u(t, z) = \sum_{j=1}^{m} \sum_{i=0}^{n-1} F_{i,j} \mathbf{1}_{(t_i, t_{i+1}]}(t) h_j(z),$$

where $0 \le t_0 < \cdots < t_n$, all $F_{i,j}$ belong to S and are $\mathcal{F}_{t_i}$-measurable, and $h_j \in C_0^1(\mathbb{R})$.

Definition 11.2.4 We define the space $\mathbb{L}_a^{1,N}$ as the space of predictable processes $u = (u(t, z))_{t \ge 0, z \in \mathbb{R}_0}$ such that, for almost all (t, z), we have that $u(t, z) \in \mathbb{D}^{1,N}$, $\partial_z u(t, z)$ exists, and

$$\|u\|_{1,N}^2 := \int_{\mathbb{R}_+} \int_{\mathbb{R}_0} \Big(E(|u(s, z)|^2) + E(|\partial_z u(s, z)|^2)$$

$$+ E(\|Du(s, z)\|_N^2) \Big) v(dz) ds < \infty.$$

The following proposition, whose proof is left as an exercise (Exercise 11.4), ensures that $\mathcal{E}_a$ is dense in $\mathbb{L}_a^{1,N}$.

Proposition 11.2.5 *The space $\mathcal{E}_a$ is dense in $\mathbb{L}_a^{1,N}$.*

If $u \in \mathbb{L}_a^{1,N}$, the stochastic integral $\int_{\mathbb{R}_+} \int_{\mathbb{R}_0} Du(t, z) \hat{N}(dt, dz)$ exists as the limit in $L^2(\Omega)$ of the stochastic integral of a sequence of approximating processes in $\mathcal{E}_a$. Moreover, we have the isometry property

$$E\left(\left\| \int_{\mathbb{R}_+} \int_{\mathbb{R}_0} Du(t, z) \hat{N}(dt, dz) \right\|_N^2 \right) = \int_{\mathbb{R}_+} \int_{\mathbb{R}_0} E(\|Du(s, z)\|_N^2) v(dz) ds.$$

$$(11.7)$$

Proposition 11.2.6 *If u belongs to $\mathbb{L}_a^{1,N}$ then the stochastic integral*

$$I := \int_{\mathbb{R}_+} \int_{\mathbb{R}_0} u(t, z) \hat{N}(dt, dz)$$

belongs to $\mathbb{D}^{1,N}$ and, for almost all $(\tau, \alpha) \in \mathbb{R}_+ \times \mathbb{R}_0$,

$$D_{\tau, \alpha} I = \partial_z u(\tau, \alpha) + \int_{\mathbb{R}_+} \int_{\mathbb{R}_0} D_{\tau, \alpha} u(t, z) \hat{N}(dt, dz). \qquad (11.8)$$

Proof By Proposition 11.2.5, the process $u \in \mathbb{L}_a^{1,N}$ can be approximated in the norm $\| \cdot \|_{1,N}$ by a sequence u^k of elementary and predictable processes of the form

$$u^k(t, z) = \sum_{j=1}^{m_k} \sum_{i=0}^{n_k-1} F_{i,j}^k \mathbf{1}_{(t_i^k, t_{i+1}^k]}(t) h_j^k(z),$$

where $F_{i,j}^k$ belongs to $\mathcal{S}$ and is $\mathcal{F}_{t_i}$-measurable and $h_j^k \in C_0^1(\mathbb{R})$. Set

$$I^k := \int_{\mathbb{R}_+} \int_{\mathbb{R}_0} u^k(t, z) \hat{N}(dt, dz).$$

Clearly, $E(|I - I^k|^2)$ converges to zero as $k \to \infty$. Then, by (11.2),

$$D_{\tau, \alpha} I^k = D_{\tau, \alpha} \left(\sum_{j=1}^{m_k} \sum_{i=0}^{n_k-1} F_{i,j}^k \int_{(t_i, t_{i+1}] \times \mathbb{R}_0} h_j^k(z) \hat{N}(dt, dz) \right)$$

$$= \int_{\mathbb{R}_+} \int_{\mathbb{R}_0} D_{\tau, \alpha} u^k(t, z) \hat{N}(dt, dz) + \partial_z u^k(\tau, \alpha).$$

We know that $E\left(\int_{\mathbb{R}_+} \int_{\mathbb{R}_0} |\partial_z u(t, z) - \partial_z u^k(t, z)|^2 \nu(dz) dt \right)$ converges to zero as $k \to \infty$. Furthermore, using the isometry property (11.7) we deduce that

$$E\left(\left\| \int_{\mathbb{R}_+} \int_{\mathbb{R}_0} (Du^k(t, z) - Du(t, z)) \hat{N}(dt, dz) \right\|_N^2 \right)$$

$$= \int_{\mathbb{R}_+} \int_{\mathbb{R}_0} E(\|Du^k(t, z) - Du(t, z)\|_N^2) \nu(dz) dt \to 0,$$

as $k \to \infty$. Therefore, $I \in \mathbb{D}^{1,N}$ and (11.8) holds. $\qquad \square$

11.3 Directional Derivative

Assume that Ω is the canonical space (9.4) associated with the Poisson random measure N, with $Z = \mathbb{R}_+ \times \mathbb{R}_0$. Remark 11.1.5 suggests defining a derivative in the direction of the jump parameter as follows. For each $\lambda \in (-1, 1)$, and any Borel set $A \subset \mathbb{R}_+ \times \mathbb{R}_0$ with $(\ell \times \nu)(A) < \infty$, we consider the measure

$$N^\lambda(A) = \int_{\mathbb{R}_+} \int_{\mathbb{R}_0} \mathbf{1}_A(s, z + \lambda) N(ds, dz)$$

and the shift θ^λ on Ω given by

$$N \circ \theta^\lambda(\omega) = N^\lambda(\omega), \quad \omega \in \Omega.$$

Set

$$y^\lambda(z) = \frac{f(z + \lambda)}{f(z)}.$$

Then, for any Borel set $A \subset \mathbb{R}_+ \times \mathbb{R}_0$ with $(\ell \times \nu)(A) < \infty$,

$$\int_{\mathbb{R}_+} \int_{\mathbb{R}_0} \mathbf{1}_A(s, z + \lambda) y^\lambda(z) \nu(dz) ds = \int_{\mathbb{R}_+} \int_{\mathbb{R}_0} \mathbf{1}_A(s, z') \nu(dz') ds.$$

Next, assume that the function f satisfies the following hypothesis:

(H2) Hypothesis (H1) holds and, for all $\lambda \in (-1, 1)$,

$$\int_{\mathbb{R}_0} (y^\lambda(z) - 1)^2 f(z) dz < \infty,$$

$$\int_{\mathbb{R}_0} \log(y^\lambda(z)) f(z) dz < \infty,$$

and

$$\int_{\mathbb{R}_0} \left(y^\lambda(z) \log(2 - y^\lambda(z)) + 1 - y^\lambda(z) \right) f(z) dz < \infty.$$

Fix $T > 0$ and let $(G_t^\lambda)_{t \in [0,T]}$ be the unique solution to the stochastic differential equation

$$dG_t^\lambda = G_{t-}^\lambda \int_{\mathbb{R}_0} (y^\lambda(z) - 1) \hat{N}(dt, dz), \quad G_0^\lambda = 1.$$

By Theorem 9.7.1, $(G_t^\lambda)_{t \in [0,T]}$ is a positive martingale and, by Itô's formula (Theorem 9.5.2),

$$G_t^\lambda = \exp \left(\int_0^t \int_{\mathbb{R}_0} \log(y^\lambda(z)) \hat{N}(ds, dz) \right.$$
$$\left. + \int_0^t \int_{\mathbb{R}_0} \left(\log(y^\lambda(z)) - y^\lambda(z) + 1 \right) \nu(dz) ds \right).$$

Furthermore, the following defines a probability measure on $(\Omega, \mathcal{F}_T)$:

$$\frac{dP^\lambda}{dP}(\omega) = G_T^\lambda(\omega), \quad \omega \in \Omega.$$

Moreover, thanks to Girsanov's theorem (Theorem 9.7.2), the law of N^λ under P^λ coincides with that of N under P.

In particular, if F is an $\mathcal{F}_T$-measurable random variable then for any bounded measurable function φ,

$$E(\varphi(F(\theta^\lambda)) G_T^\lambda) = E(\varphi(F)). \tag{11.9}$$

Therefore, if we can differentiate (11.9) with respect to λ at $\lambda = 0$, we obtain the following integration-by-parts formula.

Proposition 11.3.1 *Assume that f satisfies hypothesis* (H2) *and that*

$$\int_{\mathbb{R}_0} \frac{(f'(z))^2}{f(z)} \, dz < \infty. \tag{11.10}$$

Let F be an $\mathcal{F}_T$-measurable random variable such that, for any $\lambda \in (-1, 1)$, $F(\theta^\lambda)$ belongs to $L^2(\Omega)$, and there exists a random variable $\tilde{D}F$ in $L^2(\Omega)$ such that

$$\frac{1}{\lambda}(F(\theta^\lambda) - F) - \tilde{D}F$$

tends to zero in $L^2(\Omega)$ as $\lambda \to 0$. Then, for any function $\varphi \in C_b^1(\mathbb{R})$, the following integration-by-parts formula holds:

$$E(\varphi'(F)\tilde{D}F) = -E(\varphi(F)J_T), \tag{11.11}$$

where

$$J_T := \int_0^T \int_{\mathbb{R}_0} \frac{f'(z)}{f(z)} \hat{N}(dt, dz).$$

Proof It suffices to show that J_T is the derivative of G_T^λ with respect to λ at $\lambda = 0$ in $L^2(\Omega)$; that is, that

$$\frac{1}{\lambda}(G_T^\lambda - 1) - J_T$$

tends to zero in $L^2(\Omega)$ as $\lambda \to 0$. Indeed, this implies that we can differentiate equation (11.9) with respect to λ at $\lambda = 0$ under the expectation sign, which gives formula (11.11).

In order to prove the claim, first observe that

$$E(|G_t^\lambda - 1|^2) = \int_0^t \int_{\mathbb{R}_0} (y^\lambda(z) - 1)^2 E(|G_s^\lambda|^2) f(z) dz ds$$

$$\leq 2tg(\lambda) + 2g(\lambda) \int_0^t E(|G_s^\lambda - 1|^2) ds,$$

where $g(\lambda) := \int_{\mathbb{R}_0} (y^\lambda(z) - 1)^2 f(z) dz < \infty$. Thus, by Gronwall's lemma,

$$E(|G_t^\lambda - 1|^2) \leq 2tg(\lambda)e^{2tg(\lambda)}. \tag{11.12}$$

We now return to our goal and write

$$E(|G_T^\lambda - 1 - \lambda J_T|^2) \leq I_1 + I_2,$$

where

$$I_1 = 2g(\lambda) \int_0^T E(|G_s^\lambda - 1|^2)ds,$$

$$I_2 = 2T \int_{\mathbb{R}_0} \left(y^\lambda(z) - 1 - \lambda\frac{f'(z)}{f(z)}\right)^2 f(z)dz.$$

Using (11.12), we get

$$I_1 \le 4T^2g^2(\lambda)e^{2Tg(\lambda)}.$$

Then, using the mean-value theorem and Fatou's lemma, we obtain

$$\lim_{\lambda\to 0}\frac{g^2(\lambda)}{\lambda^2} \le \lim_{\lambda\to 0}\lambda^2\left(\int_{\mathbb{R}_0}\frac{(f'(z))^2}{f(z)}dz\right)^2 = 0,$$

which implies that

$$\lim_{\lambda\to 0}\frac{I_1}{\lambda^2} = 0.$$

For the term I_2, we again use the mean-value theorem and Fatou's lemma to conclude that

$$\lim_{\lambda\to 0}\frac{I_2}{\lambda^2} \le \int_{\mathbb{R}_0}\lim_{\lambda\to 0}\frac{(\int_0^1 f'(z + \lambda x)dx - f'(z))^2}{f(z)}dz = 0,$$

which proves the claim. □

Corollary 11.3.2 *Under the hypotheses of Proposition 11.3.1, if $\tilde{D}F \ne 0$ a.s. then the law of F is absolutely continuous with respect to the Lebesgue measure.*

Proof Let $\varphi \in C_b^\infty(\mathbb{R})$. Then, by Proposition 11.3.1,

$$|E(\varphi'(F)\tilde{D}F)| \le E(|J_T|)\|\varphi\|_\infty.$$

Therefore, by Nualart (2006, Lemma 2.1.1), the law of F under the measure $Q(d\omega) = P(d\omega)\tilde{D}F(\omega)$ is absolutely continuous with respect to the Lebesgue measure. Since $\tilde{D}F \ne 0$ a.s., Q and P are equivalent measures and the result follows. □

The next result gives the relationship between the directional derivative $\tilde{D}F$ and the Malliavin derivative DF.

Proposition 11.3.3 *For any random variable F in S,*

$$\tilde{D}F = N(DF) = \int_0^T \int_{\mathbb{R}_0} D_{t,z}FN(dt, dz), \quad a.s.$$

Proof Let $F \in S$ of the form (11.1). Then, by Definition 11.1.1, we have on the one hand

$$N(DF) = \sum_{k=1}^{m} \frac{\partial \varphi}{\partial x_k}(N(h_1), \ldots, N(h_m))N(\partial_z h_k).$$

On the other hand,

$$F(\theta^\lambda) = \varphi(N^\lambda(h_1), \ldots, N^\lambda(h_m)),$$

where $N^\lambda(h_i) = \int_0^T \int_{\mathbb{R}_0} h_i(s, z + \lambda)N(ds, dz)$. Therefore

$$\tilde{D}F = \frac{\partial}{\partial \lambda}F(\theta^\lambda)\Big|_{\lambda=0} = \sum_{k=1}^{m} \frac{\partial \varphi}{\partial x_k}(N(h_1), \ldots, N(h_m))\frac{\partial}{\partial \lambda}N^\lambda(h_k)\Big|_{\lambda=0},$$

from which the result follows, since for any $\omega \in \Omega$,

$$\frac{\partial}{\partial \lambda}N^\lambda(h_k)(\omega)\Big|_{\lambda=0} = \frac{\partial}{\partial \lambda}\omega(\theta^\lambda(h_k))\Big|_{\lambda=0} = \omega(\partial_z h_k) = N(\partial_z h_k)(\omega).$$

$\square$

11.4 Application to Diffusions with Jumps

Suppose that N is the Poisson random measure on $\mathbb{R}_+ \times \mathbb{R}_0$ associated with a pure jump Lévy process $L = (L_t)_{t \geq 0}$. Let $B = (B_t)_{t \geq 0}$ be a d-dimensional Brownian motion that is independent of L. In this section, $(\mathcal{F}_t)_{t \geq 0}$ refers to the filtration generated by the processes L and B.

We consider the set S of cylindrical random variables of the form

$$F = \varphi(B(g_1), \ldots, B(g_k), N(h_1), \ldots, N(h_m)),$$

where $\varphi \in C_0^2(\mathbb{R}^{m+k})$, $g_i \in L^2(\mathbb{R}_+; \mathbb{R}^d)$, and $h_i \in C_0^{0,2}(\mathbb{R}_+ \times \mathbb{R}_0)$.

The set S is dense in $L^2(\Omega)$. If F belongs to S, we define the Malliavin derivatives $D^B F$ and $D^N F$, considering the random variables $N(h_i)$ and $B(g_i)$, respectively, as constants. That is, for almost all $(t, z) \in \mathbb{R}_+ \times \mathbb{R}_0$ and $j = 1, \ldots, d$,

$$D_t^{B,j}F = \sum_{i=1}^{k} \frac{\partial \varphi}{\partial x_i}(B(g_1), \ldots, B(g_k), N(h_1), \ldots, N(h_m))g_i^j(t)$$

and

$$D_{t,z}^{N}F = \sum_{i=k+1}^{k+m} \frac{\partial \varphi}{\partial x_i}(B(g_1), \ldots, B(g_k), N(h_1), \ldots, N(h_m))\partial_z h_i(t, z).$$

We define the operator $L^N F$ similarly. Assume that hypothesis (H1) holds. Proposition 3.3.1 and Lemma 11.2.1 can be extended to this framework. As a consequence, the operators D^B and D^N are closable and thus they can be extended to the space $\mathcal{D}^2$ defined as the closure of $\mathcal{S}$ with respect to the seminorm

$$\||F\||_2 = \left(E(|F|^2) + E(\|D^B F\|_H^2) + E(\|D^N F\|_N^2)\right)^{1/2}.$$

Observe that $\mathcal{D}^2$ is a Hilbert space with scalar product

$$\langle F, G \rangle_{\mathcal{D}^2} = E(FG) + E(\langle D^B F, D^B G \rangle_H) + E(\langle D^N F, D^N G \rangle_N).$$

Since both derivative operators D^B and D^N satisfy the chain rule (see Propositions 3.3.2 and 11.2.2), the criterion of Proposition 7.1.2 extends to the space $\mathcal{D}^2$ as follows.

Proposition 11.4.1 *Let F be a random variable in the space $\mathcal{D}^2$ such that $\|D^B F\|_H + \|D^N F\|_N > 0$ almost surely. Then, the law of F is absolutely continuous with respect to the Lebesgue measure.*

We next consider a process $X = (X_t)_{t \geq 0}$, the solution to the following stochastic differential equation with jumps in $\mathbb{R}^m$:

$$dX_t = b(X_t)dt + \sum_{j=1}^d \sigma_j(X_t)dB_t^j + \int_{\mathbb{R}_0} c(X_{t-}, z)\hat{N}(dt, dz),$$

with initial condition $X_0 = x_0 \in \mathbb{R}^m$. The coefficients $\sigma_j, b \colon \mathbb{R}^m \to \mathbb{R}^m$ and $c \colon \mathbb{R}^m \times \mathbb{R}_0 \to \mathbb{R}^m$ are measurable functions satisfying the following Lipschitz and linear growth conditions: for all $x, y \in \mathbb{R}^m$,

$$\max_j \left(|\sigma_j(x) - \sigma_j(y)|^2, |b(x) - b(y)|^2, \int_{\mathbb{R}_0} |c(x, z) - c(y, z)|^2 \nu(dz) \right) \leq K|x - y|^2 \tag{11.13}$$

and

$$\max_j \left(|\sigma_j(x)|^2, |b(x)|^2, \int_{\mathbb{R}_0} |c(x, z)|^2 \nu(dz) \right) \leq K(1 + |x|^2). \tag{11.14}$$

The following existence and uniqueness result is well known (see e.g. Ikeda and Watanabe, 1989, Theorem 9.1).

Theorem 11.4.2 *There exists a unique càdlàg, adapted, and Markov process X on $(\Omega, \mathcal{F}, (\mathcal{F}_t)_{t \geq 0}, P)$ satisfying the integral equation*

$$X_t = x_0 + \int_0^t b(X_s)ds + \sum_{j=1}^d \int_0^t \sigma_j(X_s)dB_s^j + \int_0^t \int_{\mathbb{R}_0} c(X_{s-}, z)\hat{N}(ds, dz).$$

$$\tag{11.15}$$

Moreover, for any $T > 0$,

$$E\left(\sup_{t\in[0,T]} |X_t|^2 \right) < \infty.$$

We next give sufficient conditions on the coefficients for the solution to be differentiable in the Malliavin sense.

Theorem 11.4.3 *Assume that σ, b, and $c(\cdot, z)$ are in $C^1(\mathbb{R}^m; \mathbb{R}^m)$ for all $z \in \mathbb{R}_0$, that σ and b have bounded partial derivatives, and that, for all $x \in \mathbb{R}^m$,*

$$\int_{\mathbb{R}_0} |\partial_x c(x, z)|^2 \nu(dz) \le K.$$

Assume also that, for all $x \in \mathbb{R}^m$, $c(x, \cdot)$ belongs to $C^1(\mathbb{R}_0; \mathbb{R}^m)$ and

$$\int_{\mathbb{R}_0} |\partial_z c(x, z)|^2 \nu(dz) \le K(1 + |x|^2).$$

Then X_t^i belongs to $\mathcal{D}^2$, for all $t \ge 0$ and $i = 1, \ldots, m$, and the Malliavin derivatives $(D_r^B X_t)_{r \le t}$ and $(D_{r,\xi}^N X_t)_{r \le t, \xi \in \mathbb{R}_0}$ satisfy the following linear equations for all $j = 1, \ldots, d$:

$$D_r^{B,j} X_t = \sigma_j(X_r) + \sum_{k=1}^m \sum_{\ell=1}^d \int_r^t \partial_k \sigma_\ell(X_s) D_r^{B,j}(X_s^k) dB_s^\ell$$

$$+ \sum_{k=1}^m \int_r^t \partial_k b(X_s) D_r^{B,j}(X_s^k) ds$$

$$+ \sum_{k=1}^m \int_r^t \partial_{x_k} c(X_{s-}, z) D_r^{B,j}(X_{s-}^k) \hat{N}(ds, dz)$$

and

$$D_{r,\xi}^{N,j} X_t = \partial_z c(X_r, \xi) + \sum_{k=1}^m \sum_{\ell=1}^d \int_r^t \partial_k \sigma_\ell(X_s) D_{r,\xi}^{N,j}(X_s^k) dB_s^\ell$$

$$+ \sum_{k=1}^m \int_r^t \partial_k b(X_s) D_{r,\xi}^{N,j}(X_s^k) ds$$

$$+ \sum_{k=1}^m \int_r^t \partial_{x_k} c(X_{s-}, z) D_{r,\xi}^{N,j}(X_{s-}^k) \hat{N}(ds, dz).$$

Proof   To simplify, we assume that $b = 0$. Consider the Picard approximations given by $X_t^{(0)} = x_0$ and

$$X_t^{(n+1)} = x_0 + \sum_{j=1}^{d} \int_0^t \sigma_j(X_s^{(n)}) dB_s^j + \int_0^t \int_{\mathbb{R}_0} c(X_{s-}^{(n)}, z) \hat{N}(ds, dz),$$

if $n \geq 0$. We will prove the following claims by induction on n:

Claim 1 $X_t^{(n),i} \in \mathbb{D}^{1,2}$, for all $i = 1, \ldots, m, t \geq 0$. Moreover, for all $t \geq 0$,

$$\psi_n(t) := \sup_{0 \leq r \leq s \leq t} \mathrm{E}\left(|D_r^B X_s^{(n)}|^2\right) < \infty$$

and, for all $T > 0$ and $t \in [0, T]$,

$$\psi_{n+1}(t) \leq c_1 + c_2 \int_0^t \psi_n(s) ds,$$

for some positive constants c_1, c_2 depending on T.

Claim 2 $X_t^{(n),i} \in \mathbb{D}^{1,N}$, for all $i = 1, \ldots, m, t \geq 0$. Moreover, for all $t \geq 0$,

$$\varphi_n(t) := \sup_{s \in [0,t]} \mathrm{E}\left(\|D^N X_s^{(n)}\|_N^2\right) < \infty$$

and, for all $T > 0$ and $t \in [0, T]$,

$$\varphi_{n+1}(t) \leq c_3 + c_4 \int_0^t \varphi_n(s) ds,$$

for some positive constants c_3, c_4 depending on T.

We start by proving Claim 2. Clearly, the claim holds for $n = 0$. Assume it is true for n. First, we study the derivative with respect to N of the stochastic integral

$$G_n(t) = \int_0^t \int_{\mathbb{R}_0} c(X_s^{(n)}, z) \hat{N}(ds, dz).$$

The process $X_s^{(n)}$ is a weak version of the process $X_{s-}^{(n)}$, in the sense that $X_s^{(n)}(\omega) = X_{s-}^{(n)}(\omega)$, $d\mathrm{P}ds$-a.e. In that sense, we can write

$$\int_0^t \int_{\mathbb{R}_0} c(X_s^{(n)}, z) \hat{N}(ds, dz) = \int_0^t \int_{\mathbb{R}_0} c(X_{s-}^{(n)}, z) \hat{N}(ds, dz) \quad \text{a.e.}$$

By the chain rule (Proposition 11.2.2) and the induction hypothesis, we have that, for all $s \in [0, T]$ and $z \in \mathbb{R}_0$, $c(X_s^{(n)}, z) \in \mathbb{D}^{1,N}$ and

$$D_{r,\xi}^N c(X_s^{(n)}, z) = \sum_{k=1}^{d} \partial_{x_k} c(X_s^{(n)}, z) D_{r,\xi}^N (X_s^{(n),k}) \mathbf{1}_{\{r \leq s\}}.$$

Therefore, by the induction hypothesis and the assumptions on c, we obtain that the process $(c(X_s^{(n)}, z))_{s \in [0,T], z \in \mathbb{R}_0}$ belongs to $\mathbb{L}_a^{1,N}$. Extending Proposition 11.2.6 to an integral with respect to a Poisson random measure, we deduce that the stochastic integral $G_n(t)$ belongs to $\mathbb{D}^{1,N}$ and that

$$D_{r,\xi}^N G_n(t) = \partial_z c(X_r^{(n)}, \xi) + \sum_{k=1}^d \int_r^t \partial_{x_k} c(X_s^{(n)}, z) D_{r,\xi}^N (X_s^{(n),k}) \hat{N}(ds, dz).$$

Moreover, by (11.4), the isometry property (11.7), the induction hypothesis, and the hypotheses on c, we conclude that

$$\sup_{s \in [0,t]} \mathrm{E}\left(\|D^N G_n(s)\|_N^2\right) \leq c_1 + c_2 \int_0^t \varphi_n(s) ds.$$

The derivative with respect to N of the Itô stochastic integral follows along the same lines. This completes the proof of Claim 2.

We next prove Claim 1. Clearly, the claim holds for $n = 0$. Assume that it is true for n. We first study the derivative with respect to B of the stochastic integral $G_n(t)$. By the chain rule (Proposition 3.3.2) and the induction hypothesis, for all $s \in [0, T]$ and $z \in \mathbb{R}_0$, $c(X_s^{(n)}, z)$ belongs to $\mathbb{D}^{1,2}$ and

$$D_r^B c(X_s^{(n)}, z) = \sum_{k=1}^d \partial_{x_k} c(X_s^{(n),k}, z) D_r^B (X_s^{(n),k}) \mathbf{1}_{\{r \leq s\}}.$$

Using Proposition 3.4.3 we deduce that the random variable $G_n(t)$ belongs to $\mathbb{D}^{1,2}$ and that

$$D_r^B G_n(t) = \sum_{k=1}^d \int_r^t \partial_{x_k} c(X_{s-}^{(n)}, z) D_r^B (X_{s-}^{(n),k}) \hat{N}(ds, dz).$$

Moreover, by the isometry property of the stochastic integral (see (9.12)) and the assumption on c, we obtain

$$\sup_{0 \leq r \leq s \leq t} \mathrm{E}\left(|D_r^B G_n(s)|^2\right) \leq c_3 + c_4 \int_0^t \psi_n(s) ds.$$

The derivative with respect to B of the Itô stochastic integral follows along the same lines. This completes the proof of Claim 1. The rest of the proof follows as in Proposition 7.5.2. □

Consider now the one-dimensional version of equation (11.15), that is,

$$X_t = x_0 + \int_0^t \sigma(X_s) dB_s + \int_0^t b(X_s) ds + \int_0^t \int_{\mathbb{R}_0} c(X_{s-}, z) \hat{N}(ds, dz), \quad (11.16)$$

where $x_0 \in \mathbb{R}$ and $\sigma_j, b: \mathbb{R} \to \mathbb{R}$ and $c: \mathbb{R} \times \mathbb{R}_0 \to \mathbb{R}$ are measurable functions satisfying the Lipschitz and linear growth conditions (11.13) and (11.14), which ensure the existence of a unique càdlàg-adapted process $X = (X_t)_{t \geq 0}$ as a solution to (11.16) (see Theorem 11.4.2). Then, under the hypotheses of Theorem 11.4.3, X_t belongs to $\mathcal{D}^2$ for all $t \geq 0$, and the Malliavin derivatives $(D_r^B X_t)_{r \leq t}$ and $(D_{r,\xi}^N X_t)_{r \leq t, \xi \in \mathbb{R}_0}$ satisfy the linear equations

$$D_r^B X_t = \sigma(X_r) + \int_r^t \sigma'(X_s) D_r^B(X_s) dB_s + \int_r^t b'(X_s) D_r^B(X_s) ds$$

$$+ \int_r^t \int_{\mathbb{R}_0} \partial_x c(X_{s-}, z) D_r^B(X_s) \hat{N}(ds, dz).$$

and

$$D_{r,\xi}^N X_t = \partial_z c(X_r, \xi) + \int_r^t \sigma'(X_s) D_{r,\xi}^N(X_s) dB_s + \int_r^t b'(X_s) D_{r,\xi}^N(X_s) ds$$

$$+ \int_r^t \int_{\mathbb{R}_0} \partial_x c(X_{s-}, z) D_{r,\xi}^N(X_{s-}) \hat{N}(ds, dz).$$

We next apply a one-dimensional criterion for the existence of the density given in Proposition 11.4.1.

Theorem 11.4.4 *Let X be the unique solution to (11.16), and assume that the coefficients σ, b, and c satisfy the hypotheses of Theorem 11.4.3 and that $1 + \partial_x c(x, z) \neq 0$ for all $x \in \mathbb{R}$ and $z \in \mathbb{R}_0$. Suppose also that one of the following two conditions holds:*

(C1) *for all $x \in \mathbb{R}$, $\sigma(x) \neq 0$;*
(C2) *for all $x \in \mathbb{R}$,*

$$\int_{\mathbb{R}_0} \mathbf{1}_{\{\partial_z c(x, z) \neq 0\}} \nu(dz) = \infty.$$

Then, for all $t > 0$, X_t has a density with respect to the Lebesgue measure.

Proof By Proposition 11.4.1, it suffices to show that if either hypothesis (C1) or hypothesis (C2) holds then, for all $t > 0$ a.s.

$$\int_0^t |D_r^B X_t|^2 dr + \int_0^t \int_{\mathbb{R}_0} |D_{r,\xi}^N X_t|^2 N(dr, d\xi) > 0.$$

Observe that by Itô's formula (Theorem 9.5.2), the solutions to the linear equations satisfied by the Malliavin derivatives are given by

$$D_r^B X_t = \sigma(X_r) \exp\left(\int_r^t \sigma'(X_s)dB_s + \int_r^t (b'(X_s) - \tfrac{1}{2}(\sigma'(X_s))^2)ds \right.$$
$$+ \int_r^t \int_{\mathbb{R}_0} \log(1 + \partial_x c(X_{s-}, z))\hat{N}(ds, dz)$$
$$\left. + \int_r^t \int_{\mathbb{R}_0} (\log(1 + \partial_x c(X_{s-}, z)) - \partial_x c(X_{s-}, z))\nu(dz)ds \right).$$

and

$$D_{r,\xi}^N X_t = \partial_z c(X_{r-}, \xi) \exp\left(\int_r^t \sigma'(X_s)dB_s \right.$$
$$+ \int_r^t (b'(X_s) - \tfrac{1}{2}(\sigma'(X_s))^2)ds$$
$$+ \int_r^t \int_{\mathbb{R}_0} \log(1 + \partial_x c(X_{s-}, z))\hat{N}(ds, dz)$$
$$\left. + \int_r^t \int_{\mathbb{R}_0} (\log(1 + \partial_x c(X_{s-}, z)) - \partial_x c(X_{s-}, z))\nu(dz)ds \right).$$

We first assume that hypothesis (C1) holds. In this case, clearly, for all $t > 0$ a.s.,

$$\int_0^t |D_r^B X_t|^2 \, dr > 0.$$

Second, we assume that hypothesis (C2) holds. Then, it suffices to show that, for every $t > 0$ a.s.,

$$\int_0^t \int_{\mathbb{R}_0} \mathbf{1}_{\{\partial_z c(X_{r-}, \xi) \neq 0\}} N(dr, d\xi) > 0.$$

To this end, consider the stopping time

$$R = \inf\left\{ s > 0 : \int_0^s \int_{\mathbb{R}_0} \mathbf{1}_{\{\partial_z c(X_{r-}, \xi) \neq 0\}} N(dr, d\xi) > 0 \right\}.$$

Then it suffices to show that $R = 0$ a.s. Since X is adapted and N is a counting measure,

$$\mathrm{E}\left(\int_0^R \int_{\mathbb{R}_0} \mathbf{1}_{\{\partial_z c(X_{r-}, \xi) \neq 0\}} N(dr, d\xi) \right) = \mathrm{E}\left(\int_0^R \int_{\mathbb{R}_0} \mathbf{1}_{\{\partial_z c(X_{r-}, \xi) \neq 0\}} \nu(d\xi)dr \right) \leq 1,$$

which implies that a.s.

$$\int_0^R \int_{\mathbb{R}_0} \mathbf{1}_{\{\partial_z c(X_{r-}, \xi) \neq 0\}} \nu(d\xi)dr < \infty.$$

This contradicts hypothesis (C2) unless $R = 0$ a.s., and so the theorem is proved under hypothesis (C2). $\qquad\qquad\qquad\qquad\qquad\qquad\qquad\qquad\square$

We end this chapter by providing a different proof of Theorem 11.4.4 under hypothesis (C2) and assuming $\sigma = 0$, using the criterion of Corollary 11.3.2. See Bichteler *et al.* (1987) for its extension to the multidimensional case. Observe that, in order to apply Corollary 11.3.2, we need to assume that ν has a density satisfying the hypotheses of Proposition 11.3.1, so this alternative proof of Theorem 11.4.4 needs these additional hypotheses.

Consider the perturbed process $X_t^\lambda = X_t \circ \theta^\lambda$. Then X_t^λ satisfies the equation

$$X_t^\lambda = x_0 + \int_0^t b(X_s^\lambda)\,ds + \int_0^t \int_{\mathbb{R}_0} c(X_{s-}^\lambda, z + \lambda)\hat{N}(ds, dz)$$
$$+ \int_0^t \int_{\mathbb{R}_0} (c(X_{s-}^\lambda, z + \lambda) - c(X_{s-}^\lambda, z))\nu(dz)ds.$$

For all $\lambda \in (-1, 1)$, this equation has a unique solution in $L^2(\Omega)$. We are going to compute the process $\tilde{D}X_t$ defined in Proposition 11.3.1. Notice that the derivative $Y_t^\lambda = \partial_\lambda X_t^\lambda$ satisfies the equation

$$Y_t^\lambda = \int_0^t b'(X_s^\lambda)Y_s^\lambda ds + \int_0^t \int_{\mathbb{R}_0} \partial_x c(X_{s-}^\lambda, z + \lambda)Y_{s-}^\lambda \hat{N}(ds, dz)$$
$$+ \int_0^t \int_{\mathbb{R}_0} (\partial_x c(X_{s-}^\lambda, z + \lambda) - \partial_x c(X_{s-}^\lambda, z))Y_{s-}^\lambda \nu(dz)ds$$
$$+ \int_0^t \int_{\mathbb{R}_0} \partial_z c(X_{s-}^\lambda, z + \lambda)N(ds, dz).$$

Therefore, $\tilde{D}X_t$ satisfies the equation

$$\tilde{D}X_t = \int_0^t b'(X_s)\tilde{D}X_s ds + \int_0^t \int_{\mathbb{R}_0} \partial_x c(X_{s-}, z)\tilde{D}X_{s-}\hat{N}(ds, dz)$$
$$+ \int_0^t \int_{\mathbb{R}_0} \partial_z c(X_{s-}, z)N(ds, dz).$$

Using the method of variation of constants,

$$\tilde{D}X_t = \int_0^t \tilde{D}X_s dK_s + H_t, \qquad\qquad (11.17)$$

where

$$K_t = \int_0^t b'(X_s)ds + \int_0^t \int_{\mathbb{R}_0} \partial_x c(X_{s-}, z)\hat{N}(ds, dz)$$

and

$$H_t = \int_0^t \int_{\mathbb{R}_0} \partial_z c(X_{s-}, z) N(ds, dz).$$

By Itô's formula (Theorem 9.5.2), the solution to equation (11.17) is given by

$$\tilde{D}X_t = \mathcal{E}(K_t) \int_0^t \mathcal{E}(K_s)^{-1} \frac{\partial_z c(X_{s-}, z)}{1 + \partial_x c(X_{s-}, z)} N(ds, dz),$$

where $\mathcal{E}(K_t)$ is the Doléans–Dade exponential of K_t, given by

$$\mathcal{E}(K_t) = \exp\left(\int_0^t (b'(X_s))^2 ds + \int_0^t \int_{\mathbb{R}_0} \log(1 + \partial_x c(X_{s-}, z)) \hat{N}(ds, dz) \right.$$
$$\left. + \int_0^t \int_{\mathbb{R}_0} (\log(1 + \partial_x c(X_s, z)) - \partial_x c(X_s, z)) \nu(dz) ds \right).$$

Hypothesis (C2) implies that $\tilde{D}X_t \neq 0$ a.s., and thus Corollary 11.3.2 implies that the law of X_t has a density.

Exercises

11.1 Show formula (11.4) for any $u \in L^2(\mathcal{P})$.

11.2 Show that for any $F, G \in \mathcal{S}$, we have

$$\langle DF, DG \rangle_N = L(FG) - FLG - GLF.$$

11.3 Show Lemma 11.1.4.

11.4 Show Proposition 11.2.5.

11.5 Show the isometry property (11.7), first for elementary processes in the space $\mathcal{E}_a$ and then in the general case.

Appendix A

Basics of Stochastic Processes

A.1 Stochastic Processes

A real-valued continuous-time stochastic process is a collection of real-valued random variables $(X_t)_{t\geq 0}$ defined on a probability space $(\Omega, \mathcal{F}, P)$. The index t represents time, and one thinks of X_t as the *state* or the *position* of the process at time t. We will also consider processes with values in $\mathbb{R}^d$ (*d-dimensional processes*). In this case, for each $t \geq 0$, X_t is a d-dimensional random vector. For every fixed $\omega \in \Omega$, the mapping $t \to X_t(\omega)$ is called a *trajectory* or *sample path* of the process.

Let $(X_t)_{t\geq 0}$ be a real-valued stochastic process and let $0 \leq t_1 < \cdots < t_n$ be fixed. Then the probability distribution $P_{t_1,\ldots,t_n} = P \circ (X_{t_1}, \ldots, X_{t_n})^{-1}$ of the random vector

$$(X_{t_1}, \ldots, X_{t_n}): \Omega \to \mathbb{R}^n$$

is called the finite-dimensional marginal distribution of the process.

The following theorem establishes the existence of a stochastic process associated with a given family of finite-dimensional distributions satisfying a *consistency condition*.

Theorem A.1.1 (Kolmogorov's extension theorem) *Consider a family of probability measures*

$$\{P_{t_1,\ldots,t_n}, \ t_1 < \cdots < t_n, n \geq 1, t_i \geq 0\}$$

such that:

(i) $P_{t_1,\ldots,t_n}$ *is a probability on* $\mathbb{R}^n$;
(ii) *(consistency condition)*   *if* $\{t_{k_1} < \cdots < t_{k_m}\} \subset \{t_1 < \cdots < t_n\}$ *then* $P_{t_{k_1},\ldots,t_{k_m}}$ *is the marginal of* $P_{t_1,\ldots,t_n}$, *corresponding to the indexes* $k_1, \ldots, k_m$.

Then there exists a real-valued stochastic process $(X_t)_{t\geq 0}$ *defined in some probability space* $(\Omega, \mathcal{F}, P)$ *which has as finite-dimensional marginal distributions the family* $\{P_{t_1,...,t_n}\}$.

Example A.1.2   Let X and Y be independent random variables. Consider the stochastic process

$$X_t = tX + Y, \quad t \geq 0.$$

The sample paths of this process are lines with random coefficients. The finite-dimensional marginal distributions are given by

$$P(X_{t_1} \leq x_1, \ldots, X_{t_n} \leq x_n) = \int_{\mathbb{R}} F_X\left(\min_{1 \leq i \leq n} \frac{x_i - y}{t_i}\right) P_Y(dy),$$

where F_X denotes the cumulative distribution function of X.

A.2 Gaussian Processes

A real-valued process $(X_t)_{t \geq 0}$ is called a *second-order process* provided that $E(X_t^2) < \infty$ for all $t \geq 0$. The *mean* and the *covariance function* of a second-order process are defined by

$$m_X(t) = E(X_t),$$
$$\Gamma_X(s, t) = \text{Cov}(X_s, X_t) = E((X_s - m_X(s))(X_t - m_X(t)).$$

A real-valued stochastic process $(X_t)_{t \geq 0}$ is said to be *Gaussian* or *normal* if its finite-dimensional marginal distributions are multidimensional Gaussian laws. The mean $m_X(t)$ and the covariance function $\Gamma_X(s, t)$ of a Gaussian process determine its finite-dimensional marginal distributions. Conversely, suppose that we are given an arbitrary function $m \colon \mathbb{R}_+ \to \mathbb{R}$ and a symmetric function $\Gamma \colon \mathbb{R}_+ \times \mathbb{R}_+ \to \mathbb{R}$ which is nonnegative definite; that is,

$$\sum_{i,j=1}^{n} \Gamma(t_i, t_j) a_i a_j \geq 0$$

for all $t_i \geq 0$, $a_i \in \mathbb{R}$, and $n \geq 1$. Then there exists a Gaussian process with mean m and covariance function Γ. This result is an immediate consequence of Kolmogorov's extension theorem (Theorem A.1.1).

Example A.2.1   Let X and Y be random variables with joint Gaussian distribution. Then the process $X_t = tX + Y$, $t \geq 0$, is Gaussian with mean and covariance functions

$$m_X(t) = tE(X) + E(Y),$$
$$\Gamma_X(s, t) = st\text{Var}(X) + (s + t)\text{Cov}(X, Y) + \text{Var}(Y).$$

A.3 Equivalent Processes

Two processes, X, Y, are *equivalent* (or X is a version of Y) if, for all $t \geq 0$,

$$P(X_t = Y_t) = 1.$$

Two equivalent processes may have quite different trajectories. For example, the processes $X_t = 0$ for all $t \geq 0$ and

$$Y_t = \begin{cases} 0 & \text{if } \xi \neq t, \\ 1 & \text{if } \xi = t, \end{cases}$$

where $\xi \geq 0$ is a continuous random variable, are equivalent because $P(\xi = t) = 0$, but their trajectories are different.

Two processes X and Y are said to be *indistinguishable* if

$$X_t(\omega) = Y_t(\omega)$$

for all $t \geq 0$ and for all $\omega \in \Omega^*$, with $P(\Omega^*) = 1$. Two equivalent processes with right-continuous trajectories are indistinguishable.

A.4 Regularity of Trajectories

In order to show that a given stochastic process has continuous sample paths it is enough to have suitable estimates for the moments of the increments of the process. The following continuity criterion of Kolmogorov provides a sufficient condition of this type.

Theorem A.4.1 (Kolmogorov's continuity theorem)   *Suppose that $X = (X_t)_{t\in[0,T]}$ satisfies*

$$E(|X_t - X_s|^\beta) \leq K|t - s|^{1+\alpha},$$

for all $s, t \in [0, T]$ and for some constants $\beta, \alpha, K > 0$. Then there exists a version $\widetilde{X}$ of X such that, if $\gamma < \alpha/\beta$,

$$|\widetilde{X}_t - \widetilde{X}_s| \leq G_\gamma |t - s|^\gamma$$

for all $s, t \in [0, T]$, where G_γ is a random variable. The trajectories of $\widetilde{X}$ are Hölder continuous of order γ for any $\gamma < \alpha/\beta$.

A.5 Markov Processes

A *filtration* $(\mathcal{F}_t)_{t\geq 0}$ is an increasing family of sub-σ-fields of $\mathcal{F}$. A process $(X_t)_{t\geq 0}$ is $\mathcal{F}_t$-*adapted* if X_t is $\mathcal{F}_t$-measurable for all $t \geq 0$.

Definition A.5.1   An adapted process $(X_t)_{t\geq 0}$ is a *Markov process* with respect to a filtration $(\mathcal{F}_t)_{t\geq 0}$ if, for any $s \geq 0$, $t > 0$, and any measurable and bounded (or nonnegative) function $f: \mathbb{R} \to \mathbb{R}$,

$$E(f(X_{s+t})|\mathcal{F}_s) = E(f(X_{s+t})|X_s) \quad \text{a.s.}$$

This implies that $(X_t)_{t\geq 0}$ is also an $(\mathcal{F}_t^X)_{t\geq 0}$-Markov process, where

$$\mathcal{F}_t^X = \sigma\{X_u, 0 \leq u \leq t\}.$$

The finite-dimensional marginal distributions of a Markov process are characterized by the transition probabilities

$$p(s, x, s + t, B) = P(X_{s+t} \in B|X_s = x), \quad B \in \mathcal{B}(\mathbb{R}).$$

A.6 Stopping Times

Consider a filtration $(\mathcal{F}_t)_{t\geq 0}$ on a probability space $(\Omega, \mathcal{F}, P)$ that satisfies the following conditions:

1. if $A \in \mathcal{F}$ is such that $P(A) = 0$ then $A \in \mathcal{F}_0$;
2. the filtration is *right-continuous*; that is, for every $t \geq 0$,

$$\mathcal{F}_t = \cap_{n\geq 1}\mathcal{F}_{t+1/n}.$$

Definition A.6.1   A random variable $T: \Omega \to [0, \infty]$ is a *stopping time* with respect to this filtration if

$$\{T \leq t\} \in \mathcal{F}_t \quad \text{for all } t \geq 0.$$

The basic properties of stopping times are the following.

1. T is a stopping time if and only if $\{T < t\} \in \mathcal{F}_t$ for all $t \geq 0$.
2. $S \vee T$ and $S \wedge T$ are stopping times.
3. Given a stopping time T,

$$\mathcal{F}_T = \{A : A \cap \{T \leq t\} \in \mathcal{F}_t, \text{ for all } t \geq 0\}$$

   is a σ-field.
4. If $S \leq T$ then $\mathcal{F}_S \subset \mathcal{F}_T$.
5. Let $(X_t)_{t\geq 0}$ be a continuous and adapted process. The *hitting time* of a set $A \subset \mathbb{R}$ is defined by

$$T_A = \inf\{t \geq 0 : X_t \in A\}.$$

Then, whether A is open or closed, T_A is a stopping time.

6. Let X_t be an adapted stochastic process with right-continuous paths and let $T < \infty$ be a stopping time. Then the random variable

$$X_T(\omega) = X_{T(\omega)}(\omega)$$

is $\mathcal{F}_T$-measurable.

A.7 Martingales

Definition A.7.1 An adapted process $M = (M_t)_{t \geq 0}$ is called a *martingale* with respect to a filtration $(\mathcal{F}_t)_{t \geq 0}$ if

(i) for all $t \geq 0$, $E(|M_t|) < \infty$,
(ii) for each $s \leq t$, $E(M_t | \mathcal{F}_s) = M_s$.

Property (ii) can also be written as:

$$E(M_t - M_s | \mathcal{F}_s) = 0.$$

The process M_t is a *supermartingale* (or *submartingale*) if property (ii) is replaced by $E(M_t | \mathcal{F}_s) \leq M_s$ (or $E(M_t | \mathcal{F}_s) \geq M_s$).

The basic properties of martingales are the following.

1. For any integrable random variable X, $(E(X | \mathcal{F}_t))_{t \geq 0}$ is a martingale.
2. If M_t is a submartingale then $t \to E(M_t)$ is nondecreasing.
3. If M_t is a martingale and φ is a convex function such that $E(|\varphi(M_t)|) < \infty$ for all $t \geq 0$ then $\varphi(M_t)$ is a submartingale. In particular, if M_t is a martingale such that $E(|M_t|^p) < \infty$, for all $t \geq 0$ and for some $p \geq 1$, then $|M_t|^p$ is a submartingale.

An adapted process $(M_t)_{t \geq 0}$ is called a local martingale if there exists a sequence of stopping times $\tau_n \uparrow \infty$ such that, for each $n \geq 1$, $(M_{t \wedge \tau_n})_{t \geq 0}$ is a martingale.

The next result defines the quadratic variation of a local martingale.

Theorem A.7.2 *Let $(M_t)_{t \geq 0}$ be a continuous local martingale such that $M_0 = 0$. Let $\pi = \{0 = t_0 < t_1 < \cdots < t_n = t\}$ be a partition of the interval $[0, t]$. Then, as $|\pi| \to 0$, we have*

$$\sum_{j=0}^{n-1} (M_{t_{j+1}} - M_{t_j})^2 \xrightarrow{P} \langle M \rangle_t,$$

where the process $(\langle M \rangle_t)_{t \geq 0}$ is called the quadratic variation of the local martingale. Moreover, if $(M_t)_{t \geq 0}$ is a martingale then the convergence holds in $L^1(\Omega)$.

The quadratic variation is the unique continuous and increasing process satisfying $\langle M \rangle_0 = 0$, and the process

$$M_t^2 - \langle M \rangle_t$$

is a local martingale.

Definition A.7.3 Let $(M_t)_{t \geq 0}$ and $(N_t)_{t \geq 0}$ be two continuous local martingales such that $M_0 = N_0 = 0$. We define the *quadratic covariation* of the two local martingales as the process $(\langle M, N \rangle_t)_{t \geq 0}$ defined as

$$\langle M, N \rangle_t = \tfrac{1}{4}(\langle M + N \rangle_t - \langle M - N \rangle_t).$$

Theorem A.7.4 (Optional stopping theorem) *Suppose that $(M_t)_{t \geq 0}$ is a continuous martingale and let $S \leq T \leq K$ be two bounded stopping times. Then*

$$E(M_T | \mathcal{F}_S) = M_S.$$

This theorem implies that $E(M_T) = E(M_S)$. In the submartingale case we have $E(M_T | \mathcal{F}_S) \geq M_S$. As a consequence, if T is a bounded stopping time and $(M_t)_{t \geq 0}$ is a (sub)martingale then the process $(M_{t \wedge T})_{t \geq 0}$ is also a (sub)martingale.

Theorem A.7.5 (Doob's maximal inequalities) *Let $(M_t)_{t \in [0,T]}$ be a continuous local martingale such that $E(|M_T|^p) < \infty$ for some $p \geq 1$. Then, for all $\lambda > 0$, we have*

$$P\left(\sup_{0 \leq t \leq T} |M_t| > \lambda \right) \leq \frac{1}{\lambda^p} E(|M_T|^p). \tag{A.1}$$

If $p > 1$ then

$$E\left(\sup_{0 \leq t \leq T} |M_t|^p \right) \leq \left(\frac{p}{p-1} \right)^p E(|M_T|^p). \tag{A.2}$$

Theorem A.7.6 (Burkholder–David–Gundy inequality) *Let $(M_t)_{t \in [0,T]}$ be a continuous local martingale such that $E(|M_T|^p) < \infty$ for some $p > 0$. Then there exist constants $c(p) < C(p)$ such that*

$$c(p)E\left(\langle M \rangle_T^{p/2} \right) \leq E\left(\sup_{0 \leq t \leq T} |M_t|^p \right) \leq C(p)E\left(\langle M \rangle_T^{p/2} \right).$$

Moreover, if $(M_t)_{t \in [0,T]}$ is a continuous local martingale, with values in a separable Hilbert space H, such that $E(|M_T|^p) < \infty$ for some $p > 0$ then

$$E\left(\|M_t\|_H^p \right) \leq c_p E\left(\langle M \rangle_T^{p/2} \right), \tag{A.3}$$

where

$$\langle M \rangle_T = \sum_{i=1}^{\infty} \langle \langle M, e_i \rangle_H \rangle_T,$$

with $\{e_i, i \geq 1\}$ a complete orthonormal system in H.

References

Appelbaum, D. 2009. *Lévy Processes and Stochastic Calculus, Second Edition*. Cambridge University Press.

Bally, V., and Caramellino, L. 2011. Riesz transform and integration by parts formulas for random variables. *Stochastic Process. Appl.*, **121**, 1332–1355.

Bally, V., and Caramellino, L. 2013. Positivity and lower bounds for the density of Wiener functionals. *Potential Anal.*, **39**, 141–168.

Barlow, M. T., and Yor, M. 1981. Semi-martingale inequalities and local time. *Z. Wahrsch. Verw. Gebiete*, **55**, 237–254.

Bass, R. F. 2011. *Stochastic Processes*. Cambridge Series in Statistical and Probabilistic Mathematics, Cambridge University Press.

Baudoin, F. 2014. *Diffusion Processes and Stochastic Calculus*. Textbooks in Mathematics, European Mathematical Society.

Bichteler, K., and Jacod, J. 1983. Calcul de Malliavin pour les diffusions avec sauts, existence d'une densité dans le cas unidimensionel. In: *Proc. Séminaire de Probabilités XVII*. Volume 986 of Lecture Notes in Mathematics, pp. 132–157. Springer.

Bichteler, K., Gravereaux, J. B., and Jacod, J. 1987. *Malliavin Calculus for Processes with Jumps*. Volume 2 of Stochastic Monographs, Gordon and Breach.

Biermé, H., A., Bonami, Nourdin, I., and Peccati, G. 2012. Optimal Berry–Esseen rates on the Wiener space: the barrier of third and fourth cumulants. *ALEA Lat. Am. J. Probab. Stat.*, **9**, 473–500.

Billingsley, P. 1999. *Convergence of Probability Measures, Second Edition*. John Wiley & Sons.

Binotto, G., Nourdin, I., and Nualart, D. 2018. Weak symmetric integrals with respect to the fractional Brownian motion. *Ann. Probab.*, in press.

Bismut, J. M. 1981. Martingales, the Malliavin calculus and hypoellipticity under general Hörmander's condition. *Z. für Wahrscheinlichkeitstheorie verw. Gebiete.*, **56**, 469–505.

Bouleau, N., and Hirsch, F. 1986. Propriétés d'absolue continuité dans les espaces de Dirichlet et applications aux équations différentielles stochastiques. In: *Proc. Séminaire de Probabilités XX*. Volume 1204 of Lecture Notes in Mathematics, pp. 131–161. Springer.

Bouleau, N., and Hirsch, F. 1991. *Dirichlet Forms and Analysis on Wiener Space*. Walter de Gruyter & Co.

Breuer, P., and Major, P. 1983. Central limit theorems for nonlinear functionals of Gaussian fields. *Multivariate Anal.*, **13**, 425–441.

Brown, R. 1828. A brief description of microscopical observations made in the months of June, July and August 1827, on the particles contained in the pollen of plants; and on the general existence of active molecules in organic and inorganic bodies. *Ann. Phys.*, **14**, 294–313.

Burdzy, K., and Swanson, J. 2010. A change of variable formula with Itô correction term. *Ann. Probab.*, **38**, 1817–1869.

Cameron, R. H., and Martin, W. T. 1944. Transformation of Wiener integrals under translations. *Ann. Math.*, **45**, 386–396.

Carbery, A., and Wright, J. 2001. Distributional and L^q norm inequalities for polynomials over convex bodies in $\mathbb{R}^n$. *Math. Res. Lett.*, **8**, 233–248.

Chen, L., Goldstein, L., and Shao, Q. M. 2011. *Normal Approximation by Stein's Method*. Springer.

Chen, X., Li, W., Marcus, M. B., and Rosen, J. 2010. A CLT for the L^2 modulus of continiuty of Brownian local time. *Ann. Probab.*, **38**, 396–438.

Corcuera, J. M., and Kohatsu-Higa, A. 2011. Statistical inference and Malliavin calculus. In: *Proc. Seminar on Stochastic Analysis, Random Fields and Applications VI*, pp. 59–82. Springer.

Di Nunno, G., Oksendal, B. K., and Proske, B. 2009. *Malliavin Calculus for Lévy Processes with Applications to Finance*. Springer.

Döbler, C., and Peccati, G. 2018. The fourth moment theorem on the Poisson space. *Ann. Probab.,* in press.

Döbler, C., Vidotto, A., and Zheng, G. 2018. Fourth moment theorems on the Poisson space in any space dimension. Preprint.

Durrett, R. 2010. *Probability: Theory and Examples, Fourth Edition*. Cambridge University Press.

Dvoretzky, A., Erdös, P., and Kakutani, S. 1961. Nonincrease everywhere of the Brownian motion process. In: *Proc. 4th Berkeley Symp. Math. Statist. Probab.*, vol. 2, pp. 103–116.

Dynkin, E. B., and Yushkevich, A. A. 1956. Strong Markov processes. *Theory Prob. Appl.*, **1**, 134–139.

Einstein, A. 1905. Uber die von der molekularkinetischen Theorie der Wärme geforderte Bewegung von in ruhenden Flussigkeiten suspendierten Teilchen. *Ann. Physik*, **17**, 549–560.

Fournier, E., Lasry, J.-M., Lebuchoux, J., Lions, P.-L., and Touzi, N. 1999. Application of Malliavin calculus to Monte Carlo methods in finance. *Finance Stochastics*, **3**, 391–412.

Fournier, N. 2000. Malliavin calculus for parabolic SPDEs with jumps. *Stochastic Process. Appl.*, **87**, 115–147.

Garsia, A., Rodemich, E., and Rumsey, H. 1970/71. A real variable lemma and the continuity of paths of some Gaussian processes. *Indiana Univ. Math. J.*, **20**, 565–578.

Gaveau, B., and Trauber, P. 1982. L'intégrale stochastique comme operateur de divergence dans l'espace fonctionnel. *J. Funct. Anal.*, **46**, 230–238.

Girsanov, I. V. 1960. On transforming a certain class of stochastic processes by absolutely continuous substitution of measures. *Theory Probab. Appl.*, **5**, 285–301.

Gobet, E. 2001. Local asymptotic mixed normality property for elliptic diffusions: a Malliavin calculus approach. *Bernoulli*, **7**, 899–912.

Gobet, E. 2002. LAN property for ergodic diffusions with discrete observations. *Ann. Inst. Henri Poincaré*, **38**, 711–737.

Harnett, D., and Nualart, D. 2012. Weak convergence of the Stratonovich integral with respect to a class of Gaussian processes. *Stochastic Process. Appl.*, **122**, 3460–3505.

Harnett, D., and Nualart, D. 2013. Central limit theorem for a Stratonovich integral with Malliavin calculus. *Ann. Probab.*, **41**, 2820–2879.

Harnett, D., and Nualart, D. 2015. On Simpson's rule and fractional Brownian motion with H = 1/10. *J. Theoret. Probab.*, **28**, 1651–1688.

Hu, Y., and Nualart, D. 2005. Renormalized self-intersection local time for fractional Brownian motion. *Ann. Probab.*, **33**, 948–983.

Hu, Y., and Nualart, D. 2009. Stochastic integral representation of the L^2 modulus of continuity of Brownian local time and a central limit theorem. *Electron. Commun. Probab.*, **14**, 529–539.

Hu, Y., and Nualart, D. 2010. Central limit theorem for the third moment in space of the Brownian local time increments. *Electron. Commun. Probab.*, **15**, 396–410.

Hu, Y., Nualart, D., and Song, J. 2014a. The 4/3-variation of the derivative of the self-intersection Brownian local time and related processes. *J. Theoret. Probab.*, **27**, 789–825.

Hu, Y., Lu, F., and Nualart, D. 2014b. Convergence of densities for some functionals of Gaussian processes. *J. Funct. Anal.*, **266**, 814–875.

Hu, Y., Nualart, D., Tindel, S., and Xu, F. 2015. Density convergence in the Breuer–Major theorem for Gaussian stationary sequences. *Bernoulli*, **21**, 2336–2350.

Hu, Y., Liu, Y., and Nualart, D. 2016. Rate of convergence and asymptotic error distribution of Euler approximation schemes for fractional diffusions. *Ann. Appl. Probab.*, **26**, 1147–1207.

Hunt, G. 1956. Some theorems concerning Brownian motion. *Trans. AMS*, **81**, 294–319.

Ibragimov, I. A., and Has'minskii, R. Z. 1981. *Statistical Estimation. Asymptotic Theory*. Volume 16 of Applications of Mathematics, Springer. Translated from the Russian by Samuel Kotz.

Ikeda, N., and Watanabe, S. 1984. An introduction to Malliavin's calculus. In: *Stochastic Analysis, Proc. Taniguchi Int. Symp. on Stoch. Analysis, Katata and Kyoto, 1982*, pp. 1–52. K. Itô, Kinokuniya/North-Holland.

Ikeda, N., and Watanabe, S. 1989. *Stochastic Differential Equations and Diffusion Processes, Second Edition*. North-Holland/Kodansha.

Ishikawa, Y. 2016. *Stochastic Calculus of Variations for Jump Processes, Second Edition*. Volume 54 of De Gruyter Studies in Mathematics, De Gruyter.

Itô, K. 1944. Stochastic integral. *Proc. Imp. Acad. Tokyo*, **20**, 519–524.

Itô, K. 1951. Multiple Wiener integral. *J. Math. Soc. Japan*, **3**, 157–169.

Itô, K. 1956. Spectral type of the shift transformation of differential processes with stationary increments. *Trans. Amer. Math. Soc.*, **81**, 253–263.

Itô, K., and Nisio, M. 1968. On the convergence of sums of independent Banach space valued random variables. *Osaka J. Math.*, **5**, 35–48.

Jaramillo, A., and Nualart, D. 2018. Functional limit theorem for the self-intersection local time of the fractional Brownian motion. *Ann. Inst. Henri Poincaré*, to appear.

Kabanov, J. M. 1975. Extended stochastic integrals. *Teor. Veroyatn. Primen.*, **20**, 725–737.

Karatzas, I., and Ocone, D. 1991. A generalized Clark representation formula, with application to optimal portfolios. *Stochastics Stoch. Rep.*, **34**, 187–220.

Karatzas, I., and Shreve, S. E. 1998. *Brownian Motion and Stochastic Calculus.* Volume 113 of Graduate Texts in Mathematics, Springer.

Khinchin, A. Y. 1933. *Asymptotische Gesetze der Wahrscheinlichkeitsrechnung.* Springer.

Kim, J., and Pollard, D. 1990. Cube root asymptotics. *Ann. Statist.*, **18**, 191–219.

Kohatsu-Higa, A., and Montero, M. 2004. Malliavin calculus in finance. In: *Handbook of Computational and Numerical Methods in Finance*, pp. 111–174. Birkhäuser.

Kolmogorov, A. N. 1940. Wienersche Spiralen und einige andere interessante Kurven im Hilbertschen Raum. *C. R. (Doklady) Acad. Sci. URSS (N.S.)*, **26**, 115–118.

Kunita, H. 1984. Stochastic differential equations and stochastic flow of diffeomorphisms. In: *Proc. École d'Été de Probabilités de Saint-Flour XII, 1982.* Volume 1097 of Lecture Notes in Mathematics, pp. 144–305. Springer.

Lépingle, D., and Mémin, J. 1978. Sur l'intégrabilité des martingales exponentielles. *Z. Wahrsch. Verw. Gebiete*, **42**, 175–203.

Lévy, P. 1937. *Théorie de l'addition des variables aléatoires.* Gauthier-Villars.

Lévy, P. 1939. Sur certains processus stochastiques homogènes. *Comp. Math.*, **7**, 283–339.

Lévy, P. 1948. *Processus stochastiques et mouvement Brownien.* Gauthier-Villars.

Malliavin, P. 1978a. C^∞-hypoellipticity with degeneracy. In: *Stochastic Analysis, Proc. Int. Conf., Northwestern University, Evanston, Illinois, 1978*, pp. 199–214. Academic Press.

Malliavin, P. 1978b. C^∞-hypoellipticity with degeneracy II. In: *Stochastic Analysis, Proc. Int. Conf., Northwestern University, Evanston, Illinois, 1978*, pp. 327–340. Academic Press.

Malliavin, P. 1978c. Stochastic calculus of variations and hypoelliptic operators. In: *Proc. Int. Sym. on Stochastic Differential Equations, Kyoto, 1976*, pp. 195–263. Wiley.

Malliavin, P. 1991. *Stochastic Analysis.* Springer.

Malliavin, P., and Thalmaier, A. 2005. *Stochastic Calculus of Variations in Mathematical Finance.* Springer.

Mandelbrot, B. B., and Van Ness, J. W. 1968. Fractional Brownian motions, fractional noises and applications. *SIAM Review*, **10**, 422–437.

Meyer, P. A. 1984. Transformations de Riesz pour les lois gaussiennes. In: *Proc. Séminaire de Probabilités XVIII.* Volume 1059 of Lecture Notes in Mathematics, pp. 179–193. Springer.

Mörters, P., and Peres, Y. 2010. *Brownian Motion.* Cambridge Series in Statistical and Probabilistic Mathematics, Cambridge University Press.

Nelson, E. 1973. The free Markov field. *J. Funct. Anal.*, **12**, 217–227.

Neveu, J. 1976. Sur l'espérance conditionnelle par rapport à un mouvement Brownien. *Ann. Inst. Henri Poincaré*, **12**, 105–109.

Norris, J. 1986. Simplified Malliavin calculus. In: *Proc. Séminaire de Probabilités XX.* Volume 1204 of Lecture Notes in Mathematics, pp. 101–130. Springer.

Nourdin, I., and Nualart, D. 2010. Central limit theorems for multiple Skorokhod integrals. *J. Theoret. Probab.*, **23**, 39–64.

Nourdin, I., and Nualart, D. 2016. Fisher information and the fourth moment problem. *Ann. Inst. Henri Poincaré*, **52**, 849–867.

Nourdin, I., and Peccati, G. 2012. *Normal Approximations with Malliavin Calculus. From Stein's Method to Universality.* Cambridge University Press.

Nourdin, I., and Peccati, G. 2015. The optimal fourth moment theorem. *Proc. Amer. Math. Soc.*, **143**, 3123–3133.

Nourdin, I., and Viens, F. 2009. Density formula and concentration inequalities with Malliavin calculus. *Electron. J. Probab.*, **14**, 2287–2309.

Nourdin, I., Nualart, D., and Tudor, C. A. 2010a. Central and non-central limit theorems for weighted power variations of fractional Brownian motion. *Ann. Inst. Henri Poincaré Probab. Statist.*, **46**, 1055–1079.

Nourdin, I., Réveillac, A., and Swanson, J. 2010b. The weak Stratonovich integral with respect to fractional Brownian motion with Hurst parameter 1/6. *Electron. J. Probab.*, **15**, 2117–2162.

Nourdin, I., Nualart, D., and Peccati, G. 2016a. Quantitative stable limit theorems on the Wiener space. *Ann. Probab.*, **44**, 1–41.

Nourdin, I., Nualart, D., and Peccati, G. 2016b. Strong asymptotic independence on Wiener chaos. *Proc. Amer. Math. Soc.*, **144**, 875–886.

Novikov, A. A. 1972. On an identity for stochastic integrals. *Theory Probab. Appl.*, **17**, 717–720.

Nualart, D. 2006. *The Malliavin Calculus and Related Topics, Second Edition.* Springer.

Nualart, D., and Ortiz-Latorre, S. 2007. Central limit theorems for multiple stochastic integrals. *Stochastic Process. Appl.*, **118**, 614–628.

Nualart, D., and Pardoux, E. 1988. Stochastic calculus with anticipating integrands. *Probab. Theory Rel. Fields*, **78**, 535–581.

Nualart, D., and Peccati, G. 2005. Central limit theorems for sequences of multiple stochastic integrals. *Ann. Probab.*, **33**, 177–193.

Nualart, D., and Vives, J. 1990. Anticipative calculus for the Poisson process based on the Fock space. In: *Proc. Séminaire de Probabilités (Strasbourg)*, vol. 24, pp. 154–165.

Nualart, D., and Zakai, M. 1986. The partial Malliavin calculus. In: *Proc. Séminaire de Probabilités XVII.* Volume 1372 of Lecture Notes in Mathematics, pp. 362–381. Springer.

Ocone, D. 1984. Malliavin calculus and stochastic integral representation of diffusion processes. *Stochastics*, **12**, 161–185.

Paley, R. E. A. C., Wiener, N., and Zygmund, A. 1933. Notes on random functions. *Math. Z.*, **37**, 647–668.

Peccati, G., and Reitzner, M., eds. 2016. *Stochastic Analysis for Poisson Point Processes.* Bocconi & Springer Series.

Peccati, G., and Taqqu, M. S. 2008. Central limit theorems for double Poisson integrals. *Bernoulli*, **14**, 791–821.

Peccati, G., and Tudor, C. 2005. Gaussian limits for vector-valued multiple stochastic integrals. In: *Proc. Séminaire de Probabilités XXXVIII.* Volume 1857 of Lecture Notes in Mathematics, pp. 247–262.

Peccati, G., Solé, J. L., Taqqu, M. S., and Utzet, F. 2010. Stein's method and normal approximation of Poisson functionals. *Ann. Probab.*, **38**, 443–478.

Pipiras, V., and Taqqu, M.S. 2000. Integration questions related to fractional Brownian motion. *Probab. Theory Rel. Fields*, **118**, 251–291.

Pisier, G. 1988. Riesz transforms. A simple analytic proof of P. A. Meyer's inequality. In: *Proc. Séminaire de Probabilités XXIII*. Volume 1321 of Lecture Notes in Mathematics, pp. 485–501. Springer.

Revuz, D., and Yor, M. 1999. *Continuous Martingales and Brownian Motion*. Springer.

Rogers, L. C. G., and Walsh, J. B. 1994. The exact 4/3-variation of a process arising from Brownian motion. *Stochastics Stoch. Rep.*, **51**, 267–291.

Rosen, J. 2005. Derivatives of self-intersection local times. In: *Proc. Séminaire de Probabilités XXXVIII*. Volume 1857 of Lecture Notes in Mathematics, pp. 263–281. Springer.

Sanz-Solé, M. 2005. *Malliavin Calculus with Applications to Stochastic Partial Differential Equations*. EPFL Press.

Sato, K.-I. 1999. *Lévy Processes and Infinitely Divisible Distributions*. Volume 68 of Cambridge Studies in Advanced Mathematics, Cambridge University Press.

Segal, I. E. 1954. Abstract probability space and a theorem of Kolmogorov. *Amer. J. Math.*, **76**, 721–732.

Shigekawa, I. 1980. Derivatives of Wiener functionals and absolute continuity of induced measure. *J. Math. Kyoto Univ.*, **20**, 263–289.

Skorohod, A. V. 1975. On a generalization of a stochastic integral. *Theory Probab. Appl.*, **20**, 219–233.

Stein, C. 1986. *Approximate Computation of Expectations*. Institute of Mathematical Statistics.

Stroock, D. W. 1981a. The Malliavin calculus, a functional analytic approach. *J. Funct. Anal.*, **44**, 212–257.

Stroock, D. W. 1981b. The Malliavin calculus and its applications to second order parabolic differential equations. *Math. Syst. Theory*, **14**: 25–65, Part I; 141–171, Part II.

Stroock, D. W. 1983. Some applications of stocastic calculus to partial differential equations. In: *Proc. École d'Été de Probabilités de Saint-Flour*. Volume 976 of Lecture Notes in Mathematics, pp. 267–382. Springer.

Stroock, D. W. 1987. Homogeneous chaos revisited. In: *Proc. Séminaire de Probabilités XXI*. Volume 1247 of Lecture Notes in Mathematics, pp. 1–8. Springer.

Sugita, H. 1985. On a characterization of Wiener functionals and Malliavin's calculus. *J. Math. Kyoto Univ.*, **25**, 31–48.

Surgailis, D. 1984. On multiple Poisson stochastic integrals and associated Markov semigroups. *Probab. Math. Statist.*, **3**, 217–239.

Tanaka, H. 1963. Note on continuous additive functionals of the 1-dimensional Brownian path. *Z. Wahrscheinlichkeitstheorie*, **1**, 251–257.

Trotter, H. F. 1958. A property of Brownian motion paths. *Ill. J. Math.*, **2**, 425–433.

Wiener, N. 1923. Differential space. *J. Math. Phys.*, **2**, 131–174.

Wiener, N. 1938. The homogeneous chaos. *Amer. J. Math.*, **60**, 879–936.

Yor, M. 1986. Sur la représentation comme intégrales stochastiques des temps d'occupation du mouvement brownien dans $\mathbb{R}^d$. In: *Proc. Séminaire de Probabilités XX*. Volume 1204 of Lecture Notes in Mathematics, pp. 543–552. Springer.

Index